T0350309

Interactive Digital Television:
Technologies and Applications

George Lekakos
Athens University of Economics and Business, Greece

Konstantinos Chorianopoulos
Bauhaus University of Weimar, Germany

Georgios Doukidis
Athens University of Economics and Business, Greece

IGI PUBLISHING

Hershey • New York

Acquisition Editor:	Kristin Klinger
Senior Managing Editor:	Jennifer Neidig
Managing Editor:	Sara Reed
Assistant Managing Editor:	Sharon Berger
Development Editor:	Kristin Roth
Copy Editor:	Holly Powell
Typesetter:	Jamie Snavely
Cover Design:	Lisa Tosheff
Printed at:	Yurchak Printing Inc.

Published in the United States of America by
IGI Publishing (an imprint of IGI Global)
701 E. Chocolate Avenue
Hershey PA 17033
Tel: 717-533-8845
Fax: 717-533-8661
E-mail: cust@igi-pub.com
Web site: http://www.igi-pub.com

and in the United Kingdom by
IGI Publishing (an imprint of IGI Global)
3 Henrietta Street
Covent Garden
London WC2E 8LU
Tel: 44 20 7240 0856
Fax: 44 20 7379 0609
Web site: http://www.eurospanonline.com

Library of Congress Cataloging-in-Publication Data

Lekakos, George, 1968-
  Interactive digital television : technologies and applications / George Lekakos, Konstantinos Chorianopoulos and Georgios Doukidis.
    p. cm.
  Summary: "This book presents developments in the domain of interactive digital television covering both technical and business aspects. It focuses on analyzing concepts, research issues, and methodological approaches, analyzing existing solutions such as systems and prototypes for researchers, academicians, scholars, professionals and practitioners"--Provided by publisher.
  Includes bibliographical references and index.
  ISBN 978-1-59904-361-6 (hbk.) -- ISBN 978-1-59904-363-0 (e-book)
  1. Interactive television. 2. Digital television. 3. Internet television. I. Chorianopoulos, Konstantinos, 1975- II. Doukidis, Georgios I., 1958- III. Title.
  TK6679.3.L45 2007
  621.388'07--dc22
                                    2006039662

British Cataloguing in Publication Data
A Cataloguing in Publication record for this book is available from the British Library.

All work contributed to this book is new, previously-unpublished material. The views expressed in this book are those of the authors, but not necessarily of the publisher.

Interactive Digital Television:
Technologies and Applications

Table of Contents

Section I:
Technologies and Applications

Chapter I
Andrea Belli, Telecom Italia, Italy
Marina Geymonat, Telecom Italia, Italy
Monica Perrero, Telecom Italia, Italy
Rossana Simeoni, Telecom Italia, Italy
Monica Badella, Politecnico di Torino, Italy

Chapter II
Lyn Pemberton, University of Brighton, UK
Sanaz Fallahkhair, University of Brighton, UK

Section II:
Interaction Design

Section III:
Business and Marketing Studies

Foreword

Television is one of the most successful technological consumer products ever produced and has spread to virtually every household in Western society as well as the rest of the world. Television sets are often located in a central part of the main living room. Television viewing is a dominant part of most peoples' leisure activities and daily lives; and, for many, television has become their most important source of information and entertainment. Therefore, to say that television has a central place in our culture or that television has thoroughly changed our society and our daily lives over the past decades is an understatement that barely begins to describe reality.

Television, however, is not a static medium, neither as a technology nor as a service. Game consoles, video cassette recorders (VCRs), cable, and satellite systems have already begun to change the image of what television is and what it can be as a medium. Right now, television undergoes even more radical developments and changes with the delivery of interactive and digital TV services to the home. Terms like personal TV, customization, content-on-demand (COD), on-demand services, enhanced TV, Internet protocol TV (IPTV), SMS-TV, personal video recorders (PVR), user-created content, and more general concepts such as digitization, interactivity, convergence, cross media, the merging of television and computers, of broadcast and Internet, and so forth, point out some of the aspects involved in this process of change.

Viewing the current landscape of interactive digital television from a bird's eye perspective—there are several major trends worth noticing.

TV reception and TV content are no longer limited to the TV set, but the TV set is joined by and merged with an ever-increasing panoply of platforms and devices that can be used to receive television content and services: desktop computers, home media centers, game consoles, laptops, personal digital assistants (PDAs), mobile phones, and other wearable devices that transcend the traditional concept of TV. In this special sense, interactive digital television is television beyond TV.

One aspect of this is, on the one hand, the convergence between the world of television and the world of Telco and hence the emergence of applications that combines the strengths of interactive digital television with those of the mobile phone to create cross platform applications in the form of SMS-TV, mobile TV content, and so forth; and on the other hand the convergence between the world of television and the world of the Internet in the form of IPTV, WebTV, Web site for TV channels, TV programs, and so forth.

Television is no longer restricted to one type of technology, one type of service, and one type of application, that is, television programming, but interactive digital television is composed of different technologies and a diversity of different applications, services, and contents. TV is transformed from a mono-application to a multi-applications medium.

Television is no longer exclusively distributed from a central provider to a mass audience of viewers via broadcast. Besides cable and satellite, interactive digital television is also distributed as IPTV, as COD, or service-on-demand (SOD) upon request from the individual viewer and through peer-to-peer (P2P) networks from user to user. In this sense, interactive digital TV is TV beyond broadcast.

One aspect of this is personalized, customized, or individualized TV. In its most simple form, personalized TV is television with (PVR) functionality. With full PVR functionality the user can pause during a broadcast as content is cached on the disk and the viewer can watch the programm later. Likewise, PVR functions enable the user to rewind and fast-forward television content using the remote control. In this way, viewers can time shift the broadcast during a program, skip over commercials, and so on. In its more advanced forms personalized TV is based on models of user preferences, agents, and smart technologies. In this way, too, interactive digital TV can be seen as TV beyond broadcast.

Another aspect is enhanced TV. Enhanced TV is content—text, graphics, or video—that is overlaid on regularly displayed video content and is accessed interactively, that is, a sort of advanced teletext service or super text TV. Whether it is based on so-called local interactivity, that is, interaction between the viewer and content downloaded to the individual set-top box or based on user requests through a return channel, the enhanced part of the content is, in a way, beyond broadcast.

Interaction design in interactive digital media is no longer exclusively ruled by the usability paradigm, that is, criteria such as functionality, effectiveness, and efficiency. Looking around at contemporary activities in human computer interaction (HCI) and interaction design you are able to observe a relatively clear-defined movement from what previously used to be usability activities to what now can be defined as examinations of human experience and subjective satisfaction: Usability engineering makes way for experience design. Measurements of error rates and time per task are substituted by examinations of likeability, sociability, playability, pleasureability, and so forth. Usability experts change job titles to "user experience research managers." And former usability labs turn into human experience labs. In this sense, interaction design and experience design in relation to interactive digital television are already way beyond usability.

Television is no longer solely built on the business models of the broadcasting industry, but is also shaped by new business models and revenue opportunities based on digital and interactive capabilities as well as cross media formats. Thus, in many ways the interactive and digital television industry is on the edge of transcending the traditional broadcasting industry with profound consequences for business models, marketing, and so forth.

Television is no longer an entirely passive experience of a preplanned flow of programming with occasional zapping, but a complex combination of planned flow and interactive sequences, that is, ordinary broadcast enhanced with interactive applications, which results in a completely new media experience. In this sense, interactive digital TV is beyond TV reception.

As it appears, television—and especially digital and interactive TV—is many different things, just as the degree of interactivity varies widely. This volume deals with some of the aspects of this new situation concerning interactive digital television with a special focus on technology and applications, interaction design, and business and marketing.

Jens F. Jensen
Aalborg University, Denmark

Preface

Interactive Digital Television Definition

After a decade of interchanging enthusiasm and disappointment *interactive digital television* (iDTV) or simply *interactive TV* (iTV) enters a phase where technology developments and market conditions provide a fertile ground for its growth. The anticipated uptake of iDTV is also supported by the nature of the medium itself as a familiar and trusted medium available in almost every single household in modern societies. At the same time, the opposite is also true; TV content gradually finds its way through Internet and mobile platforms. The most important characteristic of the new medium is its ability to provide digital content enhanced with *interactive* features. Thus, besides the delivery of high quality picture and sound, iDTV promises to change the role of passive viewers of traditional analogue television by turning them into active participants of the television viewing experience.

Several definitions concerning iDTV have been proposed but little consensus has been achieved about which one depicts the characteristics and capabilities of the new medium. For example, from an engineer's point of view iTV can be regarded as digital broadcast and return channel; a content producer would refer to interactive graphics and dynamic text information; a media researcher would describe new content formats such as betting, interactive storytelling, and play-along quiz games, and a sociologist's definition would focus on the interaction between people about TV shows.

We define iTV as a user experience that involves at least one user and one or more audiovisual and networked devices. Previous definitions were focused on the technological aspects and ignored the fact that even traditional TV is potentially interactive. For example, viewers compete mentally with quiz show participants, or between colocated groups. Moreover, viewers react emotionally to TV content, they record and share TV content with friends, and they discuss shows either in real time, or afterwards. iDTV systems (such as digital video recorders, digital broadcasts, electronic program guides, Internet TV, and mobile TV) offer the potential for: (1) support for established behaviors and (2) opportunities for the emergence of new behaviors. In this sense, instead of a device, or a communication system, or a psychological/social behavior, iDTV could be defined as an experience among people, devices, and audiovisual content.

An important implication of the previous definition is that iTV applications and services are neither limited to the traditional TV device and broadcast delivery nor to the typical channels of satellite, cable, and digital terrestrial networks. In addition, alternative and complementary devices and distribution methods are considered, such as mobile phones (mobile TV), and broadband networks (DSL).

The basic household equipment of a digital television system besides the television set includes a set-top box that decodes the signal and provides additional functionality including processing and storage capabilities enabling interactive applications. The digital television signal may be delivered to households through satellite, cable, or digital terrestrial network or any other broadband IP-based network. In addition, the wide adoption of the Multimedia Home Platform (MHP) as the common open middleware platform opens new opportunities for programmers to develop sophisticated interactive applications independently from the set-top box hardware. Other important developments include standardization efforts for applications development (e.g., TV-anytime forum content metadata definition) and specifications for digital video broadcasting (DVB) though different channels (DVB-S/C/T/H specifications for broadcasting over satellite, cable, terrestrial, mobile networks), providing solutions to incompatibility and interoperability problems that have been holding back the uptake of iDTV.

Taxonomies of Interactive Applications

Interactivity, as the major feature of iDTV, can be defined as a characteristic of a medium in which the user can influence the form and/or content of the mediated presentation or experience (Ha & James, 1998; Snyder-Duch, 2001). Typically, interactivity implies two-way communication between the user and the source (broadcaster) through a return channel such as a PSTN/ISDN line plugged in a modem-enabled set-top box, or through a broadband or cable connection. Interactivity appears in various forms and levels and may be experienced even on low-end set-top boxes without storage or return channel features (called one-way interactivity or pseudo interactivity). This is achieved through the cyclic broadcast of data streams over the network (data carousel). The necessary data and code are retrieved from the network and executed on the set-top box upon user's request. Pagani (2003) classifies digital television systems according to the type of interactivity they support:

- *Diffusive systems* are those which only have one channel that runs from the information source to the user (this is known as *downstream*).
- *Interactive systems* have a return channel from the user to the information source (this is known as *upstream*).

On the other hand, Jensen (2005) suggests three iTV forms:

- *Enhanced TV.* Text, graphics, and video are broadcasted and become available to the users through interactive buttons or banners appearing on top of the normal television content.
- *Personalized TV.* The user is in control of the content through devices such as personal video/digital recorders (PVR/PDR), which can store the content and provide functionality such as pause, stop, play, rewind, forward, and so forth. In addition, advanced set-top boxes may run applications that personalize the content according to the user needs and preferences.
- *Complete iTV.* Two-way communication through a return channel. User's requests and server's responses may be synchronized or sent to the server at certain time intervals and served accordingly.

The majority of existing iTV applications is based on the enhanced-TV form. This can be attributed to a combination of factors including the limited use of return-channel enabled set-top boxes required for full interactivity options.

Interactive television users—from a uses and gratifications theoretical perspective—aim to satisfy ritualized and instrumental needs. Ritualized needs refer to a habitual use of media for time consumption and less for active goal-oriented purposes. Instrumental needs lead to a goal-oriented media usage related to information needs. In a needs-based classification of iTV applications Livaditi, Vassilopoulou, Lougos, and Chorianopoulos (2003) identify four application categories: (1) entertainment and (2) communication applications that focus on ritualized needs; (3) information and (4) transactional that focus on instrumental needs. In a recent survey in five leading—in terms of iDTV penetration—European countries (UK, Italy, Spain, France, Germany), Pagani (chapter XVIII) suggests that entertainment (53%) and information (20%) represent the majority of iTV applications. It is profound that media usage is mainly related to entertainment and therefore most of iTV applications incorporate (or should incorporate) entertainment aspects, thus blurring the distinction between application categories. Hsu, Wen, and Lee (chapter IX) identify five highly overlapping application categories (entertainment, information, education, transactions, and daily living). The latter classification extends the previous ones by the "education" category including applications such as t-learning and t-library as well as by the daily living category that embraces appliance control and t-health applications. Notably the entertainment applications are spread across all five categories. As the digital television market continues to expand and technology enables, more and more applications and application categories emerge. On the other hand, the needs-based classification suggesting the two major classes of ritualized and instrumental applications seems to be the most generic one incorporating existing taxonomies and eventually the future ones.

Interactive Applications: Opportunities and Challenges

Current and emerging applications such as video-on-demand; electronic program guide (EPG); Internet TV and mobile TV; interactive advertising; t-learning; and ambient home media applications provide opportunities and challenges for both business stakeholders and end users (viewers).

Video-on-demand changes the traditional linear viewing experience since viewers may request and receive content, which can be downloaded and stored in a PVR (such as TiVo) or an advanced set-top box. Alternatively, the viewer may pause, forward, rewind, and play causing the storage of content segments by exploiting time-shift capabilities. On the other hand, near-video-on-demand (NVoD) technology enables the selection of content that is continuously broadcasted, providing a limited spectrum of content options but with no hardware requirements (set-top box storage). The power on content provided to users poses threats and challenges to advertisers and media planners since the viewers may skip advertisements embedded in the normal program flow. This opportunity is expected to force the stakeholders involved in the advertising business to change their strategies and adjust to the new medium characteristics, finding new forms of advertising.

The vast amount of available channels turns the selection of content, relevant to the viewers' interests, a cumbersome task. EPG is one of the most widely used applications since it operates not only as a program listing service but as a portal to content-related information (known as interactive program guide [IPG]), or as an advertising message vehicle. One important challenge is the development of intelligent applications that personalize content recommending, for instance, relevant movies. The movie recommendation domain has been one of the most important research directions applying and extending recommendation methods applicable in other interactive media (e.g., Web-based recommender systems).

Television advertising over analogue television platforms has long been established as one of the most effective ways for mass-marketing activities. Digital television platforms provide new opportunities for marketers (Pramataris, Papakyriakopoulos, Lekakos, & Mylonopoulos, 2001) as they can enhance their messages with interactivity options and additional user-requested informative content enabling the collection of valuable data such as actual response rates. One of the most important consequences of interactivity is personalization of advertisements since they can be delivered to users that are most likely to respond positively to advertising messages. Among the key issues for the efficient delivery of personalized messages is the development of personalization algorithms capable of exploiting the available interaction data. The development of suitable for the domain interaction styles which in turn specifies the type of exploitable user-driven data is an important research and practical issue.

Television learning (or t-learning) is a relatively new term that emerged as a combination of typical e-learning methods applied within the iTV domain. Analogue television has been extensively used for the broadcasting of educational programs but restricting the learners' role to passive viewing. As Lytras, Lougos, and Pouloudi (2002) suggest, three domain characteristics support the learning process through television: (1) the ability to personalize content according to learner's needs and experiences, (2) digitization of content that allows for multimedia presentations of better quality of picture and sound, and (3) interactivity that gives power to users to actively participate in the learning process. However, the nature of the medium poses additional difficulties for e-learning applications. Television viewing is a group activity and mainly a content consumption process with emphasis on entertainment. The combination of television with other personal devices such as mobile phones may be considered as a gateway out of such problems, in particular for language learning applications, as Pemberton and Fallahkhair underline in chapter II where they present a learning environment incorporating TV and mobile devices for language learning.

The term *Internet TV* implies the convergence between different media-related sectors such as broadcasting, telecommunication, and information technology (IT). This convergence can be realized in different forms. On the one hand, Internet content may be accessed through television (this is also known as Web TV or Internet@TV [Jensen, 2005]) through Web browsers, or linked to iTV programs (e.g., interactive advertisements). Communication applications such as messaging, chatting, or voting during certain programs (quizzes, contests, etc.) have proven to strengthen viewer's loyalty to the specific program. However, Internet access via television may contradict with current viewing patterns. The appropriateness of uncontrolled Internet content of doubtful quality for television viewing is also questioned. Nevertheless, since Internet connection can be established via the set-top box, video content exchanges over peer-to-peer (P2P) networks seem as an attracting option for viewers and at the same time a great challenge for both broadcasters and content owners. On the other hand, the delivery of television content over IP-based platforms to personal computers, known as Internet protocol TV (IPTV), provides promising opportunities for content delivery through alternative distribution platforms. Other television content distribution platforms include third generation (3G) mobile networks featuring video-on-demand applications as well as live event broadcasts. There are still several issues to be resolved before realizing a wide acceptance of media convergence, for example, the limitations of the MHP specification to support streaming media delivery through the return channel. Demeyere, Deryckere, Ide, and Martens analyze, in chapter V, an MHP application for live-event broadcasting and video conferencing, bridging IP-based networks and DVB broadcasting.

Mobile TV applications face problems firstly related to the device characteristics (small screen, viewing angle, battery limitations) that affect the quality of video viewing. Knoche and Sasse, in chapter XIV, discuss four major requirements related to quality of service: (1) handset usability and acceptance, (2) technical performance and reliability, (3) service usability, and (4) users' satisfaction with the content. Although broadband connections enable the convergence of different types of media, one important challenge remains: the production of content that fits the different consumption models of mobile, television, and Internet content.

The convergence among different types of media and consumer electronic devices within a home network also provide the opportunity for ambient media applications. This subdomain includes home media management applications or applications that may satisfy users' needs for content consumption for entertainment/informational reasons as well as utilitarian services for safety/security and assisted living for citizens with physical inabilities. An important requirement for ambient media as a part of users' daily living is to reduce obtrusiveness and provide the means for as simple as possible interaction in particular for IT-inexperienced individuals. For example, applications based upon ambient data audio identification serving as input data without physical interaction with home devices have been proposed (Fink, Covell, & Baluja, 2006). Lugmayr, Pohl, Muehhaeuser, Kallenbach, and Chorianopoulos, in chapter VII, discuss several aspects of ambient media for home entertainment from both a technological perspective and a user-centered point of view.

Despite the aforementioned developments in the digital television domain the majority of current interactive applications are based upon "discount technology" solutions with low-end set-top boxes and the use of phone, or SMS as the return channel (Jensen, 2005). Therefore, the development of successful interactive applications—besides technological parameters—should seriously take into account the factors that affect the adoption of applications and services and the requirements stemming from end users and business stakeholders as well as the characteristics of the medium itself. More than a decade ago Lee and Lee (1995) suggested that in the design of iTV services, traditional viewing behavior should be considered avoiding extended interactivity that contradicts with the current viewing experience. Several researches agree on the profiles of iDTV interactive services adopters as young technology enthusiasts (Bjoerner, 2003; Freeman & Lessiter, 2003; Lekakos & Vrechopoulos, 2006). In the light of the recent developments in media convergence Suni in chapter XVI extends the aforementioned studies identifying specific psychological factors that drive the use of Internet television. On the other hand, Cauberghe and De Pelsmacker in chapter XVII, surveyed Belgian advertisers' perceptions on the medium to reveal that although they acknowledge its main advantages (targeting, two-way communication) they still have limited knowledge on the potential of digital television, although their attitude (intention) towards the exploitation of the medium capabilities is positive. Pagani in chapter XVIII suggests that usefulness, ease of use, and price are highly significant factors among Italians of various age groups with respect to the adoption of interactive services.

Interaction design for digital television applications is considered as one of the main research avenues in the domain, taking into account the characteristics of the medium and its differences to traditional PC-based interaction designs. In addition—as mentioned previously—it is concerned with factors (such as ease of use) that affect the adoption of iDTV applications. One important aspect of iTV applications is that they target a wide range of viewers (ultimately the analogue television viewers) representing the majority of the total population. This raises several issues concerning usability issues including the interaction of elderly or people with physical disabilities. For example, Springett and Griffiths in chapter VIII, present a number of accessibility requirements for viewers with low vision, while Iatrino and Modeo (chapter XIII) are concerned with text editing issues taking into account the limited input devices (remote control), comparing three different interaction styles. Hsu et al. (in chapter IX), Kunert and Krömker (chapter XI), and Ahonen, Turkki, Saarijärvi, Lahti, and Virtanen (chapter XII) provide solutions and guidelines for designing easy to use and useful interfaces for interactive applications. Haffner and Völkel (chapter X) discuss interaction design issues in order to support the concept of long-term relationships between broadcasters and viewers, while Knoche and Sasse (chapter XIV) provide insights concerning the quality of experience in mobile TV applications.

Book Organization

This book is concerned with certain important aspects of the iDTV domain and is divided into three main sections: Technologies and Applications (Chapters I-VII), Interaction Design (Chapters VIII-XIV), and Business and Marketing Studies (Chapters XV-XVIII). Following the aforementioned structure the book is organized in the following chapters:

Chapter I investigates current technologies and standards for convergence between the worlds of broadcasting and telecommunications. The authors propose the concept of DynamicTV as an approach that may enhance the user experience, implementing upon this concept a prototype in a DTT/IPTV environment.

Chapter II examines the potential application of iDTV to the learning of languages and describes an application that combines the strengths of iDTV with those of mobile phones to create a cross platform learning application for informal language learners.

Chapter III proposes a new hybrid approach for automatic TV content recommendation based on Semantic Web technologies, including an implementation of the approach as well as empirical evaluation results.

Chapter IV presents a model for delivering personalized ads to users while they are watching TV, which models user preferences based on characterizing not only the keywords of primary interest but also the relative weighting of those keywords.

Chapter V introduces a technology framework that can be used to add video conferencing services and live video events on the MHP, based on a bridge between IP networks and DVB broadcast channels in order to stream video that originates from an IP network into the broadcast.

Chapter VI defines a research agenda regarding the software graphics architecture for iTV acknowledging three major topics: (1) definition of a suitable declarative environment for television receivers, (2) television input (as multiple input devices) and output (multiple display devices) capabilities, and (3) models of television distribution and post-distribution.

Chapter VII deals with the development of ambient media, to satisfy the entertainment-seeking consumer. This book chapter glimpses the future of modern ambient home entertainment systems along four major lines: social implications, converging media, content, consumer, and smart devices.

Chapter VIII presents a study that examines and analyzes the performance of users with certain types of sight impairment in different interaction conditions. The outcome is the identification of issues concerning current approaches to DTV display and interactivity design as well as giving significant insights into the possible potential of and difficulties with alternative input methods.

Chapter IX proposes an activity-oriented approach to DTV user interface design, addressing DTV usefulness and usability issues. The user interface design considers both activity requirements and user requirements such as user's related product experience, mental model, and preferences.

Chapter X presents the application of concepts for long-term interaction to support long-term relationships of recipients and broadcasters in the iTV domain in contrast to classical interaction concepts that cover short-term interaction cycles. Three scenarios within the iTV domain illustrate the potential impact and long-term interaction concepts for the design of iTV applications.

Chapter XI proposes a user task-based approach to interaction design guidance for iTV applications to easily integrate with a user-centered application development process. Specific design solutions to support the user tasks "Accessing content item" and "Viewing content item" are described and empirically evaluated.

Chapter XII discusses the practical experiences of evaluating differing iTV services and proposes specific guidelines for ensuring the ease of use of interactive services. These guidelines apply to services that are transmitted in the traditional broadcast system, but they also provide a good basis for designing services in IPTV.

Chapter XIII introduces the usability problems regarding text entry using a remote control in digital terrestrial television context. It describes the comparison of three different text editing interfaces demonstrating a significant relationship between users' level of experience in text editing using mobile phone and their favorite interface.

Chapter XIV provides an overview of the key factors that influence the quality of experience (QoE) of mobile TV services. The chapter highlights the interdependencies between these factors during the delivery of content in mobile TV services to a heterogeneous set of low resolution devices.

Chapter XV examines the relation between the production for television and the Internet based on ethnographic fieldwork in two production units of the Norwegian Broadcasting Corporation (NRK). The study identifies that the two production units under consideration follow different organizational models due to differences in timing of publication on television and on the Web as well as to differences in production cycles.

Chapter XVI compares several basic statistical indicators of broadcast (traditional) television viewing and IPTV used in Estonia and shows how the structural difference between the two types of television results in different consumption models. The main conclusion is that the structure of the content to a large extent determines the uses of media.

Chapter XVII investigates the knowledge, perceptions, and intentions of 320 advertising professionals in Belgium toward the introduction and use of IDTV as a marketing communication tool. The results show that their knowledge concerning the possibilities of IDTV is very limited, but their intentions to use IDTV in the future are relatively promising.

Chapter XVIII outlines the different business models adopted in Europe in terms of contents offered and related revenue opportunities. The study addresses issues concerning how to cross the chasm of knowledge, how to explore the opportunities opened by new technologies, and which trends will influence the launch of new iTV services.

References

Bjoerner, T. (2003, April 2-4). The early interactive audience of a regional tv-station (dvb-t) in denmark. In *Proceedings of the European Conference on Interactive Television: from viewers to actors?* (pp. 91-97). Brighton, UK. Retrieved from http://www.brighton.ac.uk/interactive/euroitv/euroitv03/

Fink, M., Covell, M., & Baluja, S. (2006). Social- and interactive-television applications based on real-time ambient audio identification. In *Proceedings of the EuroITV 2006* (pp. 138-146). Athens, Greece.

Freeman, J., & Lessiter, J. (2003, April 2-4). Using attitude based segmentation to better understand viewers' usability issues with digital and interactive TV. In *Proceedings of the European Conference on Interactive Television: From viewers to actors?* (pp. 19-27). Brighton, UK. Retrieved from http://www. brighton.ac.uk/interactive/euroitv/euroitv03/

Ha, L., & James, E. L. (1998). Interactivity reexamined: A baseline analysis of early business Web sites. *Journal of Broadcasting and Electronic Media, 42*(4), 457-474.

Jensen, J. F. (2005). Interactive television: New genres, new format, new content. In *Proceedings of the Second Australasian conference on Interactive entertainment* (pp. 89-96). Sydney, Australia.

Lekakos, G., & Vrechopoulos, A. (2006). Profiling intended users of interactive and personalized digital TV advertising services in Greece. *International Journal of Internet Marketing and Advertising, 3*(3), 219-239.

Livaditi, J., Vassilopoulou, K., Lougos, C., & Chorianopoulos, K. (2003). Needs and gratifications for interactive TV applications: Implications for designers. In *Proceedings of the 36th Hawaii International Conference on System Sciences (HICSS'03)* (pp. 100b). Hawaii.

Lombard, M., & Snyder-Duch, J. (2001). Interactive advertising and presence: A framework. *Journal of Interactive Advertising, 1*(2).

Lytras, M., Lougos, C. P. C., & Pouloudi, A. (2002). Interactive television and e-learning convergence: Examining the potential of t-learning. In *Proceedings of the European Conference on E-Learning, Reading.*

Pagani, M. (2003). *Multimedia interactive digital TV: Managing the opportunities created by digital convergence.* Hershey, PA: Idea Group.

Pramataris, P., Papakyriakopoulos, D., Lekakos, G., & Mylonopoulos, N. (2001). Personalized interactive TV advertising: The Imedia business model. *Electronic Markets, 11*(1), 1-9.

Acknowledgments

The editors would like to thank the reviewers of the chapters presented in this book for the time and effort they spent to assure the quality of the manuscripts. This edited book can be considered as a result of the visions of the EuroiTV steering committee to enhance the academic's and practitioners' community in the domain of digital interactive television. The steering committees' members include, besides the editors of this book, distinguished academics, namely, Professors J. Jensen, J., Masthoff, and L. Pemberton. The birth of the idea for the publication of this book took place at the Fourth EuroiTV Conference in May 2006, organized by the Department of Management Science and Technology, Athens University of Economics and Business, Greece. Thus, the editors would like to thank the conference's host institution as well as all members of the organizing committee and colleagues for their support in the preparation of this book.

List of Reviewers

S. Agamanolis, Media Lab Europe, Ireland

L. Ardissono, University of Torino, Italy

P. Bates, pjb Associates, UK

M. Bove, MIT Media Lab, USA

A. Berglund, Linkoping University, Sweden

B. Bushoff, Sagasnet, Germany

P. Cesar, CWI, Netherlands

O. Daly-Jones, Serco Usability Services, UK

N. Ducheneaut, Palo Alto Research Center (PARC), USA

L. Eronen, Helsinky University of Technology, Finland

D. Fels, Ryerson Polytechnic University, Toronto, Canada

J. Jensen, Aalborg University, Denmark

N. Lee, ACM Computers in Entertainment

P. Looms, Danish Broadcasting (DR), Denmark

R. Luckin, University of Sussex, UK

A. Lugmayr, Tampere University of Technology, Finland

J. Masthoff, University of Aberdeen, Scotland

M. Pagani, Bocconi University, Italy

L. Pemberton, Brighton University, UK

C. Quico, TV Cabo, Portugal

B. Rao, Polytechnic University, USA

T. Rasmussen, Aalborg University, Denmark

M. Rauterberg, Technical University Eindhoven, Netherlands

B. Smith, University College Dublin, Republic of Ireland

J. Thornton, Palo Alto Research Center (PARC), USA

G. Uchyigit, Imperial College London, UK

P. Vorderer, University of South California, USA

C.Klimmt, Hanover University of Music and Drama, Germany

S. Knobloch-Westerwick, Ohio State University, USA

C. Peng, VTT, Finland,

Z. Yu, Northwestern Polytechnical University, P.R.China

L. Barkhuus, University of Glasgow, UK

J. Rode, UCI, USA

H. Knoche, UCL, UK

R. Mandler, ABC, USA

A. Dollar, ITV Alliance, USA

Section I

Technologies and Applications

Chapter I

DynamicTV:
The Long Tail Applied to Broadband-Broadcast Integration

Andrea Belli, Telecom Italia, Italy

Marina Geymonat, Telecom Italia, Italy

Monica Perrero, Telecom Italia, Italy

Rossana Simeoni, Telecom Italia, Italy

Monica Badella, Politecnico di Torino, Italy

Abstract

This chapter proposes a way to exploit the ongoing convergence between the worlds of broadcasting and Telco in order to find an added value for the audience. We propose DynamicTV, a concept based on the idea that this scenario may enhance the user TV-watching experience by providing not only a more comprehensive choice (content accessible via the broadband channel are virtually infinite) but most important, a way to take the user from a pure passive TV consumption to a more interactive, on demand, nonlinear experience. The maturity of the current technologies and standards for convergence has been investigated by implementing a prototype in a digital terrestrial television (DTT)/Internet protocol television (IPTV) environment of the DynamicTV concept. Our conclusion is that both standards and technologies need some effort in order to allow a real service of this kind to be offered on the market.

Introduction

In the beginning of television, starting with a single channel available only certain hours a day, the technical solution behind broadcasting shaped the whole TV idea into a simple "the broadcaster chooses, for all the audience, a few things to watch" and that was it. Some choice was given to the audience when it became possible to select among several channels.

Then the digital television came with digital video broadcasting (DVB)-T/S/C set-top boxes, thus increasing the number of available channels to choose from; then a small modem-based return channel appeared on some STBs, allowing some kind of interactivity; but still the idea behind it was that the broadcasters chose the content to propose to their big audience.

Currently, the existing DVB-T interactive applications are generally limited to easy graphic electronic program guides (EPGs) with some hyper-textual information and few T-government solutions, created and promoted in order to try and decrease the digital divide. The availability of a broadband IP access in the DVB-T/S STBs can bring things one step further: not only we can watch on TV what the broadcaster sends over the air, but we are now able to reach multimedia content available anywhere on the Web and watch it on the TV set, without anyone choosing for us what to watch.

But this may not yet be what we want. We already know the drawbacks of this Internet-based approach and we know how little it fits with the idea of watching TV: first you have to *know* what you want to see. Then you have to *find* it. And who knows what you end up with: There is no control, no catalog, no quality check for Internet content. And you must be lucky to find the content in the *coding* supported by your TV set. And at a *bit rate* compatible with the broadband access line that serves you, provided that your network provider has properly dimensioned the network. These are just a few of the parameters that need to be checked in order to be able to enjoy Internet content on TV. Though the availability of an enormous amount of content and the absence of any form of censorship might be appealing, however, experience with the Web shows how rarely things work properly, thus allowing a relaxed TV-like experience.

What's next then? The next step along this path is IPTV. The network infrastructure needed to support this kind of broadcasting service is in fact automatically adequate for on-demand content distribution, too. This means that without any significant investment, the broadcaster can complement its business with a content-on-demand offer being delivered on exactly the same network as the broadcasted content.

This actually changes the "we choose, for all of you, a few things to watch" model. There is still an intermediation component, since the content must be checked against public morality; coded according to the desired quality; stored; tagged with metadata and possibly DRM-ed; and so forth, but the range of choices for the audience is potentially infinite. The new model could then become more like *"we propose a few things to all of you, and we also make available a huge quantity of content (fit to your morality and your TV!) among which you may find almost anything you like."*

This is what we start to call the *convergence* between broadband and broadcast, and this is the end of the tale and the beginning of our studies, described in this chapter with the following structure:

- The background section explains how the *long tail approach* could be applied to the TV world.
- We propose a concept relying on broadband-broadcast convergence.
- We then describe the prototype we developed, which implements a part of the concept.
- Finally, we derive the conclusions and highlight the main items that qualify for further research.

Background

A broadband IP return channel added to a DVB-T STB or IPTV architecture may, if properly implemented in the network, for example:

- Adding *unicast* distribution to the audience based on single requests;
- Providing TV quality through the IP network and on the STB, enhance the TV watching experience by:
- Overcoming OTA frequency limitation and increasing the number of TV channels thanks to multicast streaming and broadband ADSL access.
- Making any kind of content available on demand to a single customer, accessing both Web sites and servers over the Internet (with the drawbacks we stated in the previous section) or a large repository of TV-formatted content (made available over the IPTV network architecture by a provider).

The questions are now the following:

Question no. 1—How may these enhancements be translated into something actually appealing for the audience?

We try and answer this question by applying the well known long tail concept (Anderson, 2004) to the interactive TV (iTV) world (either as DVB-T with the broadband return channel, or as IPTV).

As reported by Anderson (2004), studies on TV audiences show that people make a massive consumption of few television channels. This means that 80% of the audience watches about 20% of the available content, which may not necessarily be the best, the most interesting, or the ones of highest artistic value.

The advent of the iTV world gives the opportunity to make a large amount of multimedia content available to the whole TV audience. This is a significant step, since it makes the so called *niche content*, previously accessible only to a little number of people, largely available.

Moreover it is important to highlight that viewers' behavior changed in these years. Based on some trial experiences carried out in our labs,[1] we can state that the number of people who usually watch TV alone is increasing (Bonaventura et al., 2004). Families generally have more than one television, and the habit of watching TV alone is gradually increasing.

On the other hand, the rising tide of available content creates the need of new tools for guiding users to find what is of interest to them.

There is the need not only *to filter* that amount of information, to present it on the basis of the viewers' *actual interests* but also to find a way—and this is something new—to choose and propose content according to the viewer's *mood* at that very moment. The term *mood* refers to a feeling state that is pervasive, temporary (unlike attitudes or temperaments that are enduring), and without an object focus (unlike emotions that are related to a particular object) (Hess, Kacen, & Kim, 2005). So, next to the traditional recommendation techniques (based on user profiling) our aim is to study a way to include the viewers' mood as a parameter based on which to choose and offer them the most suitable content. Obviously, the problem of finding methods and techniques (e.g., sensor networks) for detecting a viewer's mood will also be covered in the future.

Besides accessing niche content, viewers might have the opportunity to navigate among content and easily create dynamic, often nonlinear solutions, choosing among the entire spectrum of available content. There is in fact a number of interesting works more related to a *channel-centric approach* (Chorianopoulos, 2004; Rowson, Gossweiler, & MacDonald, 2005), which we do not cover in this study. We focus instead on other possible ways to guide the viewer into the entire spectrum of available content, starting from the widely used PC-search approach, bearing in mind that the TV set is *not* a PC and therefore "Googling it!" might not be the right answer.

Let us see why: based on some trial experience carried out in our labs it appears that, in order to give to this new iTV approach the chance to be successful, it is important not to twist the traditional "couch potato" audience approach to television (Bonaventura et al., 2004).

When we talk about a passive TV consumption we refer only to the fact that, in the traditional TV context, audience consumes broadcast content in a pure "PUSH" mode, without explicitly influencing the programming or interacting with the medium, apart from zapping.

Although the Internet is both the repository of the content and the distribution network, the experience we are trying to give to our customers is not a PC-like experience moved into the TV, but rather a television-enhanced one, taking advantage of the new technology. In this scenario, a blank page with a square where you can type something might not be exactly what we are looking for. Unluckily, there are currently many iTV applications designed according to PC interface metaphors, not matching audience habits and creating difficulties, for example, in reading lots of text or using too technical devices. These are the issues we considered when studying the DynamicTV solution.

Question no.2—Are broadcasters and Telcos really converging?

Nowadays in Europe there is still a clear separation between broadcasters and Telco, regarding both the technical aspects of STBs and the creation and offering of hybrid services. Furthermore, today's technology and standards do not yet provide a good framework to manage an integrated solution.

Despite the work going on in the DVB Forum, both in the multimedia home platform (MHP) group for defining the DVB-hyper text markup language (HTML) and in the DVB-IPI group for Service Discovery and Signalling, the current commercial IPTV platforms are still proprietary. This makes integration with DTT particularly difficult.

On the other hand, new reasons are rising to fulfil interoperability between the two worlds, witnessed by a few newborn initiatives:

- The diffusion of personal video recorders (PVR) (Looms, 2005), among which TiVo (http://www.tivo.com) represents one of the most popular examples, which amplifies the use of *time shifting* and allows a *nonlinear consumption* of television, shows the audience is getting used (and they like it!) to a more personalized way of watching TV where every single user may choose when to watch things. Above all, users can decide what to see and what to skip: This is jeopardizing the broadcasters' traditional advertising model.
- Recently, solutions that integrate broadcast and broadband worlds have been prototyped or just offered. Examples are NTT (Kawazoe & Hayama, 2004) and Comcast (Baumgartner, 2005) solutions that link and propose, during a live sequence of an event, on-demand content related to that event.

Technical developments on the network side (triple-play solutions) and on the set-top box side (hybrid DTT-IPTV or DTT-broadband boxes) make the broadband-broadcast interoperability a reality for the near future and a very interesting opportunity.

DynamicTV: A Concept of iTV

To explore this new universe we have started with studies on the standard situation, the technological challenges, and the user experience needs. Next, we have tried to go deep into the problems by designing and implementing a small prototype that fits into the big picture, aiming at exemplifying as clearly as possible a little part of it.

In this section we describe the service concept called DynamicTV addressing broadband and broadcast integration, while in section 4 we describe the prototype implementation of a part of it. This experience proved to be very useful to highlight the levels of maturity in technology and standards.

Broadcast tends to be associated with the transmission of live content, while broadband is associated with making content available on demand to the users. The idea of DynamicTV is based on the following observation: when a user switches the TV on, a common habit is to zap the available channels and see what is going on in each of them; at the end, the user would most probably end up on the channel broadcasting the content nearest to his/her mood at that moment. At this point, a number of different situations may occur (see Table 1): The idea is that for each situation there is something useful or appealing that may be offered to that very user, based on the broadband capability.

The idea is that on-demand content can be promoted, somehow piggybacking the broadcasted content and then delivered via broadband network (Flores, 2005). It is possible with the currently available technology to associate video on demand (VoD) content to live TV and to propose it to the audience, thus obtaining a non-linear TV experience.

The challenge here is to capture audience attitude and to offer them exactly what they may appreciate in that moment. The choice can be made based on what they have just seen, liked, or disliked. We can say that the proposals we offer to the user are somehow based on a *mood management concept*. People watching television generally want to satisfy a specific need that depends on their current mood (Bonnici, 2003; Gentikow, 2005; Taylor & Harper, 2002).

Table 1 shows a list of service functions coming from different application fields: TV (EPG), Internet search, e-commerce services (proposals), and DVD (in depth and special content); an iTV application could group together all these features. In fact, each of these functions could satisfy a specific need a viewer could have

Table 1. Relationship between the mood of the viewer and the TV service offered

Mood during consumption	Service function
Generic interest	EPG
Specific interest	Search
Trust in proposal	Proposals
High interest while watching	In depth
High interest after watching	Special content

while watching TV. This table has been studied by a group of researchers in our labs and it can be a starting point to define new ways of TV interaction and to build new interfaces that allow a simpler and more intuitive access to a large number of contents (Chorianopoulos & Spinellis, 2004).

As can be seen in Table 1, one of these moods corresponds with the EPG service function. In this case the overloaded word "EPG" means something more than a simple schedule of the broadcasters' offer to the audience. This "more" can be declined in many different ways, starting from the idea of an integrated DTT+Broadband EPG, to a personalized EPG (Ardissono et al., 2003), to an interactive program guide (IPG) allowing the user to request on-demand content "related" to the standard schedule, to any other feature that might be thought of.

If the user has a specific interest, he/she would like to find—in a very quick way—the TV content he/she wants to watch. We stated in the previous section the reasons why we believe it is important not to adopt on TV the PC-search approach as it is, because it could decrease usability.

First of all the user could have some difficulties in writing the search term because they do not have a keyboard. So, it might be an interesting challenge to think of other mechanisms that allow the user to write in a quick and easy way, on a TV set. Moreover, the search results cannot be displayed as a simple list: It might be necessary to conceive a new way of showing results on TV (using, for example, some kind of filter).

Instead of the EPG and search, which have largely been discussed in other papers, we have focused on the last three lines of Table 1. They can be explained as follows:

- **Trust in proposal:** The viewer is currently watching something that is near to his/her mood, but it is not exactly what he/she is interested in (e.g., one of your favorite movies, which you have already seen three times, including special content and backstage; a movie that you do not like but that features your favorite actor; a movie which is "act 2" of something you have not seen yet, etc.). In this case the user should be able to ask for "proposals" to the provider he/she trusts. The content proposed should encounter his/her expectations and should be available on demand with a click (they are delivered unicast on the broadband infrastructure).

- **High interest while watching—In depth:** The viewer is interested in what he/she is watching and he/she wants to know more about it, *while* carrying on watching. This further information (and possibly more content related to it) is delivered unicast on the broadband network and might invite the viewer to interact with the program.

- **High interest after watching—Special content:** In this case, we assume that the viewer has just finished watching a content of interest and wants to have more of it, exactly as it happens with the "special content" on a DVD, with the advantage that content can be continuously updated, enriched, and modified. This is also the way to promote related "content" such as merchandising, soundtracks, ring tones, wallpapers, and so forth. In this case, broadband is not only a way to deliver content and information, but might also be a way to allow billing for merchandising.

In all the previous cases, the current program on the broadcast channel is a good context in which to propose *other* content, that would otherwise be unknown or difficult to access (niche content), with a high probability of the user liking them (Flores, 2005). In that way DynamicTV applies the long tail approach using live programs (hits) as a portal to access niche content.

Starting from these considerations, DynamicTV tries to be a comprehensive concept, covering all the aforementioned moods, and therefore able to satisfy most of the expectations a viewer might have.

There are a lot of technologies needed for the implementation of the DynamicTV concept, and most of them are involved even if one tries to implement only a subset of the whole big picture. At least, this is what we have experienced by implementing a part of the concept as a prototype.

The Prototype

User Experience

The first prototype of DynamicTV assumes a user watching a program (e.g., *The Matrix*) broadcasted on the DTT. While the viewer is watching the movie, he/she can start the interactive application by selecting the red button on the bottom of the screen. As soon as the application is entered, a horizontal menu appears, offering the five services represented in Table 1, the last three of which are the interesting ones for this chapter:

- EPG shows the programming for all the different channels.
- Ricerca (Search) enables the user to look for a particular content. The paradigm to be used is still under study.
- Proposte (Proposals) shows a list of content (movies, in this case) somehow similar to the current movie.
- Approfondimenti (In depth) presents a set of information regarding the movie that the viewer is watching.
- Contenuti speciali (Special content) presents some multimedia content related to the movie (see Figure 1).

If the viewer wants to get some additional information about the current movie, he/she can select the "In depth" item. A new page will be shown with information about the movie, such as: genre, actors, director, synopsis, awards, and so on. This information is sent unicast only to that viewer via the broadband network. While reading this information, the viewer can carry on watching the program which is still visible in a corner of the screen (see Figure 2).

At the end of the movie (or at any other time) the viewer can choose the "Contenuti Speciali" item and watch some "special content" (as with a DVD), like, for example, the making of the movie, backstage, interviews, music videos, and so forth. Content are sent unicast only to that viewer via the broadband network.

Figure 1. Screenshot of the menu page

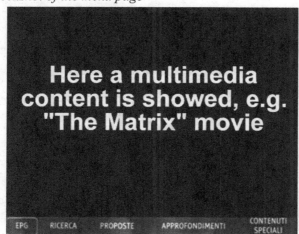

Figure 2. Screenshot of approfondimenti (in depth), showed on a user STB

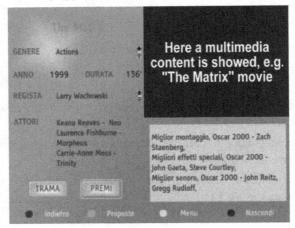

If the user is interested in receiving "proposals" about something else to watch, somehow similar or related to *The Matrix,* he/she can choose the "Proposte" item. The DTT application shows the viewer a list of movies related to *The Matrix* by genre, actors, or director. The list is retrieved on a server via the broadband network. More recommendation criteria should be used in order to enhance the effectiveness of the suggestions (collaborative filtering, context awareness, etc.), and they will be one of our main subjects for research in the near future.

Figure 3. Screenshot of "proposte" (proposals) suggested on the basis of what the user is watching

If the viewer selects one of the recommended movies, the application shows him/ her a page with some additional information about it. At this point the viewer can request that movie and watch it *instead* of the current program. The selected movie will be streamed unicast via the broadband network (see Figure 3).

Standards Adoption

The underlying idea that stands behind the concept is that the features we have described should be available to *all* households, independently of the specific technology people have bought and the provider they have chosen. The only way leading to such a situation is to rely on standards.

We have already reported on lack of standards for IPTV platforms; this is one of the reasons why it has been decided to prototype the concept by implementing an interactive MHP application. The other main reason standing behind this MHP choice is the unique situation of the Italian market where 3.5 Million DTT set-top boxes with MHP have already been deployed.

A starting point for broadband-broadcast integration might be identified in a DTT STB equipped with a broadband return channel. Through the return channel it is possible to retrieve information and content stored on servers in the network and offer them to the user. Though MHP is a mature standard, the streaming via the broadband channel is still needing some standardization effort.

Our interactive application has been developed using DTTRun, an authoring tool created in our labs that enables us to build MHP applications with a graphic tool. Initially the prototype was run on a PC, using the Reference Implementation MHP 1.0.3 provided by IRT, using the ethernet return channel for the VoD streaming. Later the application has been run on the Access Media hybrid set-top box IT.BOX, mounting an MHP stack provided by Alticast. With the proprietary application program interfaces (APIs), kindly provided for research purposes by Access Media, it has been possible to stream Motion Picture Experts Group (MPEG)-2 videos via the broadband return channel.

TV-Anytime

The metadata used to describe content and to make suggestions are based on TV-anytime (TVA) metadata specification (www.tv-anytime.org). One of the objectives of these specifications is to enable applications to exploit local persistent storage in consumer electronics platforms in a network-independent way; the applications would therefore work in the same way on various content delivery mechanisms (e.g., ATSC, DVB, DBS, and others), as well as the on the Internet and enhanced TV.

TVA specifications include more than metadata: They are designed for interoperable and integrated systems and cover the entire content provision cycle, from content creators, through service providers, to the consumers, providing also necessary security structures to protect the interests of all parties involved. TVA Forum closely collaborates with the Pro-MPEG Forum and the advanced authoring format (AAF) Association, and builds on top of the widely accepted MPEG standards; for example, TVA specifications use the MPEG-7 description definition language (DLL) to describe metadata structure based on the extensible markup language (XML) encoding of metadata, as well as MPEG-7 classification schemes.

Through the metadata schema definition it is possible to indicate information about general aspects such as title, genre, synopsis, and so on. Moreover, it is possible to insert information about temporal and spatial location of content as well as about the user. The adoption of TVA tags allows us to describe TV program schedules and represent program descriptions.

TVA taxonomy is used as the basis for the recommender system to provide proposals based on correlations between metadata about the available content. TVA standards have also been adopted by the British Broadcasting Corporation (BBC) and NTT to maintain information about their programming.

Prototype Architecture

DynamicTV architecture is based on role separation between the broadcast and the broadband side, which are linked only at set-top box level. The broadcaster is responsible for the service transmission over the air, consisting of a DVB-T transport stream with an associated MHP application, thus the broadcaster knows what program a user is watching at the moment the application is run (see Figure 4).

On the broadband side, a network provider hosts the DynamicTV engine server application which is in charge of answering the invoked requests, through the return channel, by the MHP application running on the set-top box.

The DynamicTV engine queries a TVA database containing metadata about programs and content, encapsulates the response in XML format and delivers it to the client application running on the set-top box. The structure of the database reflects the TVA standard taxonomy.

For the client application, the DynamicTV engine represents the "door" to reach the world of broadband services (in our case, the VoD service). The engine server exposes Web services which can be invoked by the client either through simple object access protocol (SOAP) over hyper text transfer protocol (HTTP) or simple HTTP requests.

It is important to remark that Web services are the core components of the overall architecture since they are accessible from any network-enabled device, allowing

Figure 4. DynamicTV prototype architecture

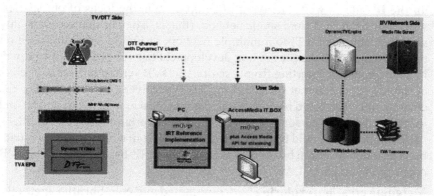

us to port DynamicTV service model to different architectures including the IPTV environment.

Future Trends

From the technical point of view, thanks to the implementation of a small prototype as realistic as possible in all its components, we have been able to identify challenges and limitations that such concept brings up: We have brought into perspective how much the current DTT and IPTV architectures, hybrid set-top box maturity, availability of metadata, and standardization activities need to be enhanced to make the foundation for a new TV service.

Regarding hybrid terminals, given that the cross-over point between MPEG-2 video transmission bandwidths and sustained broadband access bandwidths in volume deployments has now been reached, the IPTV service market will be enabled soon by "hybrid" set-top boxes, equipment that support both legacy TV services (such as cable and satellite) and IP-based services; the sad part of the story is that at the moment most set-top box vendors only propose what can be defined as "combo" prototypes: An MHP stack and the IPTV functionality are mounted on the same equipment, but no interoperability between the two worlds is provided. In this scenario it is not easy to implement a network independent, content-based service, which is necessary in order to allow the user to forget about what channel (IPTV or DTT) the multimedia content is coming from.

The next steps of our research will therefore focus on the identified limitations, trying particularly to overcome the following:

- **Application level interoperability not only in the MHP environment but on the IP side as well.** The types of equipment, in the focus of our studies, will allow us to shift from single network (that is, a device that supports only satellite, cable, or DTT) to a multiple network or hybrid solution; based on these assumptions, the set-top box development has the potential for growth in many directions, starting from a stationary DVB set-top box with wireless or cable-bound ethernet interface and DVB receiver (DVB-S(2)/T/C), the ethernet interface could be connected to home local area network (LAN) and exchange data with personal devices on home networks (e.g., DLNA profiles), MHP applications could be deployed on the box from a broadcast or a broadband channel or set-top boxes may as well receive IP-based content via DVB connection and home LAN.

- **Metadata creation, exploring both automated and collaborative methods.** In the next years, more and more content producers and broadcasters should add metadata information to the content they publish. Unfortunately, there will probably always be a large amount of content not carrying any metadata or with very little or no associated information. We are studying two different strategies to overcome these problems:

  - **On one hand metadata could be retrieved automatically using analysis of texts associated to the audio-video content and/or extracting its low-level features (e.g., color, frames, etc.)** and possibly combining them into high-level features with a stronger semantic value. This last technique is a challenging longer-term research subject.

  - **On the other hand, metadata could be created not only by content producers**, but also by end users, a kind of TV Wikipedia, where users can actively contribute to the collaborative creation and the editing of content information.

- **Personal content availability on TV, by means of mediated infrastructure promoting and protecting personal content.** Thanks to the Internet evolution the gap between content producer and end user has significantly decreased. The birth of personal sites, blogs, forums, and so forth allows every single user to publish self-made content in a very simple and cheap way, potentially making it available to a large number of people. Until some years ago self-made content were mostly texts (articles, blogs), but during the last years people publishing their own pictures, music, and videos are increasing; the diffusion of videoblogs is an example of this trend. The progressive proliferation of cheap and easy-to-use digital cameras and the diffusion of mobile phones with integrated cameras that allow users to take pictures and to record a video, allow any person to become "director" of their own productions or, using a recent definition, a *prosumer.*[2] All these elements, together with the access bandwith increase, make it possible for everyone to create, edit, and publish audio/video content.

The aim of DynamicTV is to extend the prosumer role also to the TV scenario. From the role of simple viewer any user is given the opportunity to have an active "production" role, with the advantage that the content proposed may potentially reach a broader audience thanks to the DynamicTV model.

- **Effective paradigm for this new TV, designed on purpose rather than borrowed from the PC.** One of the most important challenges of DynamicTV is to propose to the user a new way to use television. The aim is not only to transform the typical *passive* user experience in a more interactive one, but also to create an innovative metaphor able to help the user to find content he/she really likes in a simple, non-intrusive, and even entertaining way.

The process to create a new TV interaction paradigm has to take into account some main issues. First of all, with the introduction of IPTV and DTT, the number of available content enormously increased, creating the need for an evolution of the classical TV paradigms, zapping and EPG. Besides, as we already stated, we want to avoid applying to the TV a PC-like approach; currently some IPTV implementations use a *portal-like interface*, but this might not be the best choice: TV screens have a lower resolution than PCs, and users usually watch TV from a farther distance so it is not possible to show too much text on the screen or to use small fonts. Even if in the near future, a high definition TV with a large screen will spread, reading a lot of text on TV will always be a difficult and probably not entertaining task.

These considerations do not mean that the current approaches are bad, but we believe that it is time to identify the next TV interaction paradigm, able to mix in an innovative way the best aspects related to both media, to create a really new and relaxing experience that may differ significantly from the current one. To make this transformation easier it may be interesting to introduce new devices, other than the remote control, to simplify some new tasks such as, for example, text writing (to search a movie or to log on to a service).

Conclusion

The main ideas that have been presented in this paper can be summarized as follows:

- The concept of DynamicTV proposes a way to apply the concept of long tail to the TV world, exploiting:
  ° The broadcast-broadband convergence that might happen in the future.
  ° The fact that hit content can be used as a way to propose and promote niche content somehow related to them.

- The greatest challenge for this idea to be successful is to be able to lead gently and smoothly the TV audience from a passive model of content consumption to a more interactive experience, without the viewers even being aware of the change in attitude.

References

Anderson, C. (2004). The long tail. Forget squeezing millions from a few megahits at the top of the charts. The future of entertainment is in the millions of niche markets at the shallow end of bitstream. *Wired Magazine, 12*(10), 170-177.

Ardissono, L., Portis, F., Torasso, P., Bellifemine, F., Chiarotto, A., & Difino, A. (2001). *Architecture of a system for the generation of personalized Electronic Program Guides.* Paper presented at UM2001 Workshop on Personalization in Future TV, Sonthofen, Germany.

Baumgartner, J. (2005, April 4). Comcast guides rich media IPG. *Communication Engineering & Design Magazine.* Retrieved November 27, 2005, from http://www.cedmagazine.com/article/CA6264482.html

Bonaventura, S., Zanassa, I., Bellavita, S., Menegotto, P., Lisa, S., & Geymonat, M. (2004). *Trial BHL: Risultati del focus group sulla "Casa connessa al mondo."*

Bonnici, S. (2003). Which channel is that on? A design model for electronic programme guides. In *Proceedings of the 1ˢᵗ European Conference on Interactive Television: from Viewers to Actors?* Brighton, UK.

Chorianopoulos, K. (2004). *Virtual television channels: Conceptual model, user interface design and affective usability evaluation.* Unpublished doctoral dissertation, Athens University of Economics and Business, Greece.

Chorianopoulos, K., & Spinellis, D. (2004). Affective usability evaluation for an interactive music television channel. *Computers in Entertainment, 2*(3), 14.

Flores, N. (2005). *Broadcast as promoter of on-demand.* On demand media Weblog. Retrieved November 27, 2005, from http://ondemandmedia.typepad.com/odm/2005/09/broadcast_as_pr.html

Gentikow, B. (2005). *Limiting factors for embracing ITV: The embodiment of media use.* Paper presented at EuroITV 2005, 3ʳᵈ. European Conference on Interactive Television.

Hess, J., Kacen, J., & Kim, J. (2005). *Mood management dynamics: Fitting differential equations to consumer panel data.* Paper presented at Marketing dynamics conference, University of California, Davis.

Kawazoe, K., & Hayama, S. (2004). *Cooperative broadcasting and communication technology for a new way of viewing TV. NTT Technical Review Online, 8,* 50. Retrieved November 27, 2005, from http://www.ntt.co.jp/tr/0408/-files/ntr200408050.pdf

Looms, P. O. (2005, March). *PVRs and the free-to-air television market in Europe.* Paper presented at EuroITV2005, 3rd European Conference on Interactive Television.

Rowson, J. A., Gossweiler, R., & MacDonald, K. (2005, March). *PHIZ: Discovering TVs long tail through a channel—Centric model.* Paper presented at EuroITV 2005, 3rd European Conference on Interactive Television.

Taylor, A. S., & Harper, R. (2002). Switching on to switch off: An analysis of routine TV watching habits and their implications for electronic programme guide design. *usableiTV, 1*(3), 7-13.

Endnotes

[1]   This trial was structured in 10 focus group and involved 68 people.

[2]   Prosumer is a term formed by contracting either the word producer or professional with the word consumer (from Wikipedia http://www.wikipedia.org)

Chapter II

Interactive Television as a Vehicle for Language Learning

Lyn Pemberton, University of Brighton, UK

Sanaz Fallahkhair, University of Brighton, UK

Abstract

Designers of technological support for learning have an ever-increasing selection of platforms and devices at their disposal, as desktop computers are joined by mobile phones, personal digital assistants (PDAs), laptops, games consoles, and wearable devices. Digital television is a recent addition to this panoply of platforms and devices. Television has always had educational aims and effects, and it is natural for digital television to continue this tendency. In this chapter we examine the potential application of interactive digital television to the learning of languages. We explore the "fit" between the capabilities of the medium and the requirements of language learners. We then describe an application that combines the strengths of interactive digital television with those of the mobile phone to create a cross-platform learning application for informal language learners.[1]

Introduction

New media technologies seem inevitably destined to be hailed as the breakthrough for language learning. In particular technologies that allow interactivity—language labs, audiocassettes, videodiscs of various formats, and the multiplicity of interactive applications hosted on the Web have all excited the interest of language learners and teachers. Currently, interactive mobile technologies are being harnessed to support many applications for learning—formal and informal; for children and adults; in classrooms and out on field trips; at home; and on the move. Broadcast technologies, that is, radio and television, have not been so central. Until comparatively recently, these were more difficult to integrate into the classroom experience, needing careful timing and recording. In addition the opportunities for interactive engagement with broadcast media are less obvious than for e-mail, chat, and so on. Moreover, television is overwhelmingly seen as a leisure technology. This has meant that compared with other media and platforms, television has been rather under-used in formal language teaching environments.[2] In this chapter we discuss the potential of broadcast and interactive television (iTV) for language learning. We then describe an attempt to integrate support for language learning into broadcast television, with the additional support of a mobile phone application.

Television and Language Learning

Outside the classroom, conventional television is already a powerful learning tool for language learners. There is a wealth of anecdotal evidence of individuals learning other languages from exposure to televised soap operas, films, and sports reporting. Television offers a rich multimedia experience, where learners can immerse themselves in materials from the target language and culture. Nakhimovsky (1997), quoted in Underwood (2002) notes, perhaps *"of all subjects, foreign language instruction can benefit from multimedia materials most obviously ... the most difficult task facing a language instructor is to show the deep semantic and cultural differences hidden behind dictionary equivalents ... the movie and its script makes that task much easier ... students can see the clash between their expectations and the realia of a different culture."* This material, unlike some of the stilted exchanges found in textbooks, tends to be worth watching in its own right, with up-to-date, ever-changing content displaying a range of speakers and contexts. In its non-interactive state it is an excellent medium for learners to practice comprehension skills and also to acquire background cultural knowledge (Sherington, 1973). Many television shows constitute important cultural events providing a shared reference for people sharing or aspiring to share a culture, and for many language learners, watching, understanding, and enjoying TV and films in the foreign language can itself be a learning objective (Liontas, 2002).

Subtitling is used in conventional television to support viewing in other languages and also in the guise of closed captions to help viewers with hearing problems. Several projects have analysed the use of TV with first and second language subtitles as an aid to comprehension, retention of second language vocabulary, and improving reading skills. Studies suggest that even watching TV with an L2 audio track and L1 subtitling could lead to incidental second language learning (Koolstra & Beentjes, 1999; Meinhof, 1998). Koskinen, Knable, Markham, Jensema, and Kane (1996) suggest that captioned television can be used as an effective instructional tool in learning vocabulary and concepts. In a study of the effect of captioned television on incidental vocabulary acquisition by adult English as a second language (ESL) learners, they assessed vocabulary knowledge in viewers who watched TV with and without captions, identifying "*a statistically significant difference in favour of captioned TV*" (p. 368).

Interactive digital television (IDTV) offers the opportunity to go further with subtitling, captioning, and other language support facilities (Atwere & Bates, 2000; Pemberton, 2002; Underwood, 2002). However, its potential is not widely exploited, as current levels of interactivity are relatively limited, constrained by the components of the IDTV set up, namely the set-top box and its software, the on-screen display, and the remote control (Gawlinski, 2003). Generic interactive facilities of interest to language learners include:

- Choosing among subtitling or captioning options
- Viewing supplementary information on screen—to access before, during, or after a broadcast (known as *enhanced TV*)
- Accessing Web pages on TV
- Selecting from alternative audio and video streams (e.g., choosing a scientific or generalist commentary stream)
- Using communication tools such as chat and e-mail

IDTV's strengths, then, are those of conventional TV—provision of current, culturally relevant video and audio in a leisure setting—plus additional, currently restricted, hypermedia and communication facilities. This is particularly important for providers aiming to extend learning opportunities to non-classroom settings, for example, traditionally hard-to-reach groups, such as care givers and the disabled, as well as time-poor professionals. For the first group, alternative learning facilities via the Internet or face-to-face classes may not be an option, while for the latter group, time for dedicated learning activities is limited. Learning via a familiar technology as a side effect of a leisure activity should be attractive to many.[3]

Some of these strengths were confirmed by focus groups we carried out with adult learners of English and other foreign languages (see Fallahkhair, Masthoff, & Pemberton, 2004, for a fuller account). Our results indicated that language learners do not perceive iTV as a medium for formal learning, but as a form of entertainment that may have the side effect of incidental learning. Even our most fanatical language learners were not keen to watch TV programmes specifically made for the language student. In addition, they were aware of the tensions that imposing specifically educational material might have on their fellow viewers. However, the up-to-date authentic material broadcast on TV was very attractive to them, and they perceived it as bringing many valuable learning opportunities. Hence, rather than creating iTV programmes specifically for language learning, developers looking to exploit the learning potential of iTV for language learners should add interactive enhancements to existing, engaging programmes, supporting informal rather than formal learning via programmes the viewer might watch spontaneously even without language learning opportunities. They should also provide support for viewers. Our participants appreciated any support that helped them obtain more from their foreign language viewing. In particular multimedia presentations of material, with different media complementing each other and providing context, were seen to facilitate understanding: subtitles made it easier to follow rapid speech, gestures, and other graphical information expressed extra-linguistic meaning, a visual setting anchored the meaning of spoken language, and so on. IDTV could scaffold understanding even further by providing a selection of levels of support in appropriate complementary media, either through the television screen or via a separate device such as the mobile phone.

Television has its drawbacks as a technology for language learners. Broadcast TV is (currently) schedule-bound, non-interruptible, and often perceived by learners as just too fast (Broady, 1997; Fallahkhair, Pemberton, & Masthoff, 2004). If a learner is watching alone, using a personal video recorder (PVR) can remedy this, but at the cost of "liveness," which may be important in watching some programmes such as a sports fixture or a quiz show with its own interactive services. Television viewing is often a group activity however, and replay via a PVR would be unpopular in this situation. In addition, unlike, say, the mobile phone, television is not normally perceived as a personal companion technology, and of course it is not designed for "any place, any time" interaction. It may be that commercial attempts to deliver TV to the PC and mobile phones will eventually capture the imagination of users, but for the moment our model of television is the traditional one of the TV as a "digital hearth" (Moores, 1996).

It would be unrealistic to expect a single technology to give equal support for the multiple types of cognitive activity that are involved in language learning (Pemberton, 2002). In order to overcome some of the characteristics that act as drawbacks in iTV, it may be possible to use other technologies—the Internet, interactive video, mobile technologies—to fill some gaps. Mobile phones in particular have been in focus over the last 3 years and provide characteristics that provide the flexible, personal, discreet, anytime-anywhere functionality identified as lacking in iTV.

Mobile Technologies and Language Learning

The potential value of learning via mobile devices or m-learning has been widely realised (Leung & Chan, 2003; Naismith, Lonsdale, Vavoula, & Sharples, 2005; Sharples, 2000). The mobile phone is a particularly interesting mobile technology for learning technologists as it has been very rapidly adopted by non-technically minded users who would not have ready access to laptops or PDAs and has woven itself into the fabric of their everyday life. The mobile phone has the potential to enhance learning experiences by providing two-way voice communication, by allowing access to learning materials outside fixed times and places and by sup-porting information retrieval and capture on the move, now including multimedia information. While many researchers have seen the benefits of mobile devices as a supplement to classroom teaching in formal school, college, and training contexts, our main interest is in individual learners in informal (i.e., not necessarily curricu-lum-bound) settings. Here another characteristic of the mobile phone comes into play, its status as a personal companion device. For language learning in particular this is a valuable characteristic. Second language learners are often found with a dog-eared pocket dictionary or a personal vocabulary book about their person, and the mobile phone can serve as a very natural digital equivalent of these personal technologies.

Several researchers have begun to investigate the potential of mobile devices for language teaching (Godwin-Jones, 2004; Kadyte, 2003; Tan & Liu, 2004). While many of these projects are oriented towards formal education, others target inde-pendent adult learners. One project that successfully exploits some of the strengths of the mobile phone is the INLET project (Pincas, 2004), which developed an in-novative mobile phone support system to encourage tourists to learn some Greek at the Athens Olympic Games. The system provided a number of facilities for learning useful Greek phrases in various categories such as "basic" (e.g., greeting, numbers, basic words), "where" (e.g., phrases for asking direction, going by bus, taxi, and trains), "when" (e.g., asking times, today, now, tomorrow), "Olympic Sport" (game name, athletics, fencing, etc.) and "buying" (asking price, money, expressions like expensive, cheap, etc.). Users, typically recruited at Athens airport, were able to register for free SMS messages containing useful phrases to be sent to their mobile phones each day. They could also request on the spot SMS translations from other languages into Greek.

Projects like this use mobile phones to retrieve information about individual vocabulary items and to interrogate information sources about their current context. For instance, a spectator at the Olympic Games might send an SMS asking how to say "pole vault" in Greek during the field events. Other aspects of mobile phones have proved less popular with general users and tend not to be exploited in research projects. Despite the prevalence of SMS messaging, lack of good input devices tends to restrict the length of text entries on mobile phones, so there is little point in designing a mobile application for extensive text composition. Mobiles also tend to not be used for display of long passages of text for reading, probably because of the small screen size, and although advanced phones are now capable of displaying rich multimedia, video viewing on mobiles has not been widely popular and is therefore probably not a fruitful path to follow in educational projects.

TAMALLE Project: Functionality

The TV and mobile assisted language learning environment (TAMALLE) project was originally planned as a way of testing out the potential of iTV for language learning. As a result of both our own reflection on the capabilities of the platform and also of feedback from focus group studies with independent adult language learners, we decided instead on a dual device solution combining iTV and a mobile phone. The aim of the system is to support advanced learners of English as a second language in their television viewing, as just one element in their language learning activities. These would typically be recent immigrants, foreign students, asylum seekers, and so on, though we would expect some native speakers with low literacy skills also to benefit. As the focus of the learners will be on media consumption rather than on conscious language learning, this support is designed to be as discreet and non-intrusive as possible. The system provides support, in the form of captions and other on-screen displays, for comprehension of specific language (or sometimes cultural) items for viewers as they watch English language programmes. These items can be incorporated by learners into their personal *learning sphere*, a private data storage area, which is accessible both via the TV and on their mobile phone. The mobile can further support learners' understanding of the programme by enabling them to access the summary of programmes as well as difficult language and cultural items that may appear in a programme. These language items can be accessed prior to, during, and after the show. Viewers are also able to add, search for, and remove items from/into their personal spheres. Even without television, the mobile service is useful for learning the new recommended language items and as a tool for managing personal knowledge.

The following is a brief overview of the system's functionality:

Figure 1. Supporting comprehension of difficult language items across iTV and mobile phones

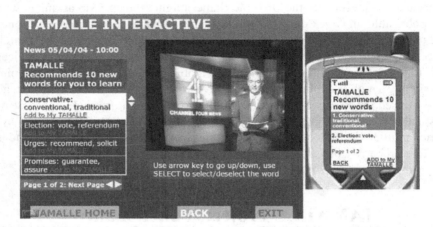

- **Just-in-time comprehension support:** The "Words in Action" function provides textual annotation similar to subtitles on the television screen. The individual items may explicate a word (e.g., Tory = Conservative) or identify a scene or individual (10 Downing St = the British Prime Minister's residence). Items are selected on the basis of rules devised as a result of studies with language learners. These indicated that a number of categories, including geographical names, terms relating narrowly to Western and UK culture, slang and prepositional verbs would be useful. The design locates the call-to-action dialogue on the iTV side rather than the mobile phone since this just-in-time support will only be beneficial during the programme show time and not before or after. However, if the learner prefers not to display annotations on the TV screen, perhaps to avoid inconvenience to others or embarrassment to himself, the "Words in Action" content can be delivered to their mobile phone in synchrony with the programme.

- **Recommended language items:** Difficult or unusual language items from the dialogue or commentary will also be transcribed for TAMALLE viewers in a static list. Viewers who are logged in may select "Recommended Words" to see a list of language items with explanation, which can also be added to their personal learning sphere. The service is also accessible via mobile phone.

- **Supporting overall understanding:** The viewer's overall understanding may be improved by having access to a summary of programme content. This will differ according to genre, with the news being summarised as headlines, a drama as a brief plot summary, and so on. This is augmented by an on-screen learner dictionary, also available on the mobile phone.

- **Managing personal learning sphere:** The system enables learners to manage their personal learning sphere via both iTV and mobile interfaces. The recommended words can be added to a personal vocabulary list for later practice. Learners can view all their saved language items from the main menu. They can also search for specific language items and remove those no longer wanted.

Architecture

TAMALLE is enabled via a learning management system located in the broadcast end or back-end tier. This provides content to both set-top box and mobile devices and also holds learning content or learning objects in a database on the back-end tier (MySQL). In the front-end tier we have the set-top box and WAP-enabled mobile devices. Two way communications can be established between set top box and back-end tier through telephone modem, ADSL, or broadband cable, while mobile phone devices communicate with the back-end tier through the WAP protocol. For interactive SMS messaging we use an SMS gateway provider, in our case the UK SMS2mail provider (for details see Fallahkhair, 2004). This architecture is sketched in Figure 2.

Figure 2. TAMALLE dual device architecture

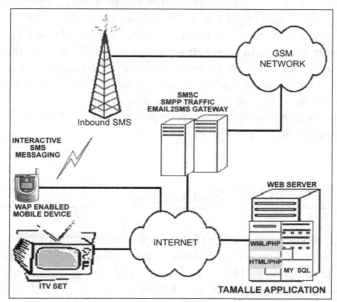

Interaction Design Issues

The design of TAMALLE attempts to make best use of the strengths of each device and to distribute functionality and displays appropriately across them. However, another design aim was to provide consistency in terms of look (structure, icons, text) and feel (navigation, interaction) across both devices in order to guarantee the usability of the overall system. Many functions can be carried out via either interface, and it would be too much of a burden on the user to expect them to learn different methods for each device (Robertson, Wharton, Ashworth, & Franske, 1996).

The physical characteristics and limitations of the mobile phone and the iTV system such as screen size, resolution, and memory capacity, place heavy constraints on the possible display options and user interactions, which come as a shock to developers used to the relative freedom of PC and Web environments. The typical contexts of use for each device are a source of further constraints: The TV remote has to be usable while watching TV, with as few distracting "head down" moments as possible, while mobile phones are very often used while on the move or in combination with other tasks. In addition, relatively strong conventions have quickly grown up among users and manufacturers of these devices, and designers have to be careful to adhere to platform guidelines so as to avoid upsetting users' expectations (or in the case of iTV applications, having their service rejected by platform providers). Unlike a PC or Web service, which will probably be the user's entire focus of attention, TAMALLE will be just one among many others—the user will expect them to behave roughly the same as other services on the device.

The conventions of iTV have developed quite independently of those for mobile services, a situation which might cause difficulties for a cross-platform design. On the one hand, the system needs internal consistency if it is to be usable, learnable, and recognisable as the same application though on a different platform. On the other hand, an iTV service must be consistent with iTV design guidelines if it is to be usable and learnable, and similarly for a mobile service. The implications of these constraints on major aspects of interaction design, namely, screen design, navigation, interaction, and branding are explored in more detail in Fallahkhair, Pemberton, and Griffiths (2005).

Evaluation

The prototype version of the system has been evaluated with potential users. Usability evaluation for interactive systems is now a well-established part of any user-centred design process. However, new types of *beyond-the-desktop* systems, such as mobile devices, wearables, and leisure technologies involve factors that usability-oriented techniques cannot capture. Like cross-platform design, cross-platform evaluation is a much more complex challenge than evaluation of desktop applications, with both iTV and mobile phones bringing difficulties in their own right (Isomursu, Kuutti, & Väinämö, 2004; Pemberton & Griffiths, 2003). We used a multi-method evaluation regime requiring potential users to engage in activities to gauge usability, perceived usefulness, and desirability. In order to cover these aspects, a combination of observations, questionnaires, interviews, and card sorting was used over sessions that lasted around 2 hours for each participant. Fourteen sessions were carried out, with members of the target user group, that is, advanced learners of English as a foreign language.

Conventional Usability Observation

The evaluation sessions took place once the system was a functioning prototype. The sessions took place in a specialist usability lab, a room furnished as a domestic lounge, where participants could be observed using television via a one-way window and remotely controlled cameras. After a brief screening questionnaire and an introduction to the technology, the participants worked through two relatively well-specified scenarios. This enabled participants to be in a position to give informed responses in the subsequent phases as well as supplying detailed usability data. As other researchers have discovered, "... *(m)otivation is a major issue. Instructions and tasks need to be very simple and readily achievable*" (Daly-Jones, 2002, p. 3). The facilitator therefore played a rather active role in this part of the session, encouraging the participant at points when otherwise they might have given up. This obviously has the disadvantage of interrupting the natural flow of events, but was felt to be unavoidable (Pemberton & Griffiths, 2003, p. 883).

Usability Questionnaire

For another view of usability as experienced by the participants, we used a usability questionnaire based on ISO 9241 (Gediga, Hamborg, & Doentsch, 2000). This consisted of a selection of 54 questions in six areas of usability:

- Conformity with user expectations
- Self-descriptiveness
- Controllability
- Learnability
- Suitability for learning
- Error tolerance

The answers were on a scale of 1-5. TAMALLE scored an overall 3.9, giving a global measure of its current usability to weigh against more detailed evidence from observations of ease of use of particular functions.

Feature Rating

We were very interested in capturing users' perceptions of the usefulness and possible effectiveness of the TAMALLE system's services. Participants were asked to rate the usefulness of 21 features on scale of 1 (useless) to 5 (useful). For instance, they were asked to rate the provision of the programme summary on a mobile phone, saving words in personal vocabulary space and so on. In addition they were asked for their three favourite features and the three least liked, with explanations of their choices. Participants also answered questions on their perceptions of the particular language skills that might be supported by TAMALLE (i.e., spoken or written comprehension or production) and the TV genres that might best lend themselves to augmentation via TAMALLE services.

These questions elicited different reactions from the usability-oriented phase of the evaluation, and it was clear that participants were judging potential efficacy in supporting their language skills rather than ease of use.

Desirability Rating

The next phase of the evaluation sessions aimed to gauge the overall acceptability of the proposed system. Both iTV and mobile phone use are, in their different ways, highly personal, and we were particularly interested in finding out whether users would find the language support services intrusive and inappropriate. If this were so, then usability and likely efficacy might be irrelevant. We used the Product Reaction Card method developed by (Benedek & Miner, 2002). Users are presented with 118 cards marked with a range of adjectives showing judgments of the system, for example, annoying, appealing, personal, powerful, valuable, inconsistent, and so on. Users select a collection of relevant terms and then refine their selection to give a top five of most appropriate judgements, to which they can add comments.

To conclude this phase we also asked users about overall attitudes to the software; its perceived strengths and weaknesses; and the likelihood of them using it if it were to become commercially available.

Although the application of several techniques in a single session was quite demanding of participants' time and concentration, the sessions succeeded in gathering a rich range of data, evaluating the usability, the perceived efficacy, and the desirability of the TAMALLE system. Reactions were very encouraging, with many of the participants indicating that they would be likely to use such a service and find it useful. The results can now be integrated into the next version of the system. It should be pointed out, though, that as a learning support system its effectiveness can only be evaluated when it is deployed over an extended period in a realistic setting.

Further Work

There are many further issues that remain unresolved for the TAMALLE project. One of these is related to learning content. Textbooks, language lab programmes, and interactive videos can be handcrafted. However, if annotated mainstream broadcast material is to be produced in a sustainable way, a large measure of automation will be needed. To this end we intend to extend the user test programme to develop robust rules for selecting language items for annotation.

We are also investigating the most successful format for summaries for broadcast material in different genres. Other areas for further work include the addition of manual input to the personal learning sphere, use of voice output, and the possibilities for adaptivity and personalisation.

Acknowledgment

We would like to acknowledge the contribution of our colleague Judith Masthoff, particularly in the requirements stage, plus the members of our various user groups for their contributions. Thanks also to the EuroITV reviewers for useful feedback.

References

Atwere, D., & Bates, P. J. (2003). *Interactive TV: A learning platform with potential.* Learning and Skills Development Agency. Retrieved from http://www.lsda. org.uk/files/PDF/1443.pdf

Benedek, J., & Miner, T. (2002). Measuring desirability: New methods for evaluating desirability in a usability lab setting. In *Proceedings of Usability Professionals' Association 2002*, Orlando, FL. Retrieved from www.microsoft. com/usability/publications.mspx

Broady, E. (1997). Old technology, new technology: Video makes a come-back. In A. Korsvold & B. Ruschoff (Eds.), *New technologies in language learning and teaching*. Council of Europe.

Daly-Jones, O. (2002). Navigating your TV: The usability of electronic programme guides. *Usable iTV, 1/3,* 2-6.

Fallahkhair, S. (2004). Media convergence: An architecture for iTV and mobile phone based interactive language learning. In *Proceedings of EuroITV 2004, Brighton* (pp. 177-182).

Fallahkhair, S., Masthoff, J., & Pemberton, L. (2004). Learning languages from interactive television: Language learners reflect on techniques and technologies. In L. Cantoni & C. McLoughlin (Eds.), *World Conference on Educational Multimedia, Hypermedia & Telecommunications EdMedia 2004* (pp. 4336-4343).

Fallahkhair, S., Pemberton, L., & Griffiths, R. N. (2005). Dual device user interface design for ubiquitous language learning: Mobile phone and interactive television (iTV). *IEEE International Conference on Wireless and Mobile Technology for Education* Tokushima, Japan.

Fallahkhair, S., Pemberton, L., & Masthoff, J. (2004). A dual device scenario for informal language learning: Interactive television meets the mobile phone. In Kinshuk, Looi, C., E. Sutinen, D. Sampson, I. Aedo, L. Uden, & E. Kahkonen (Eds.), *The 4th IEEE International Conference on Advanced Learning Technologies (ICALT 2004)* (pp.16-20).

Gawlinksi, M. (2003). *Interactive television production*. London: Focal Press.

Gediga, G., Hamborg, K. C., & Doentsch, I. (2000). The isometrics usability inventory: An operationalisation of ISO 9241/10. *Behaviour and Information Technology, 18,* 151-164.

Godwin-Jones, B. (2004). Emerging technologies language in action: From Webquest to virtual realities. *Language Learning and Technology.* 8(3).

Isomursu, M., Kuutti, K., & Väinämö, S. (2004). Experience clip: Method for user participation and evaluation of mobile concepts. In *Proceedings of the 8th Conference on Participatory Design* (pp. 83-92). Toronto.

Kadyte, V. (2003). Learning can happen anywhere: A mobile system for language learning. In J. Attwell & J. Savill-Smith (Eds.), *MLearn: Learning with mobile devices, research and development*. Learning and Skills Development Agency.

Koolstra, C. M., & Beentjes, J.W.J. (1999). Children's vocabulary acquisition in a foreign language through watching subtitled television programs at home. *Educational Technology Research and Development, 47*(1), 51-60.

Koskinen, P., Knable, J., Markham, P., Jensema, C., & W. Kane, K. (1996). Captioned television and the vocabulary acquisition of adult second language correctional facility residents. *Journal of Educational Technology Systems, 24*(4), 359-373.

Leung, C., & Chan, Y. (2003). Mobile learning: A new paradigm in electronic learning. In *Proceeding of 3rd IEEE International Conference on Advanced Learning Technology.*

Liontas, J. I. (2002). CALLMedia digital technology: Whither in the new millenium? *CALICO Journal, 19*(2), 315-330.

Meinhof, U. H. (1998). *Language learning in the age of satellite television.* UK: Oxford University Press.

Moores, S. (1996). *Satellite television and everyday life.* UK: University of Luton Press.

Naismith, L., Lonsdale, P., Vavoula, G., & Sharples, M. (2005). *Literature review in mobile technologies and learning,* UK: University of Bristol, Nesta Futurelab.

Nakhimovsky, A. (1997). A multimedia authoring tool for language instruction: Interactions of pedagogy and design. *Journal of Educational Computing Research, 17*(3), 261-274.

Pemberton, L. (2002). The potential of interactive television for delivering individualised language learning. In *Proceedings of workshop on Future TV,* ITS 2002, San Sebastian.

Pemberton, L., & Griffiths, R. N. (2003). Usability evaluation techniques for interactive television. In *Proceedings of HCI International,* Crete, 2004.

Pincas, A. (2004). Using mobile support for use of Greek during the Olympic Games 2004. In *Proceeding of M-learn Conference* 2004.

Purushotma, R. (2005, January). You're not studying, you're just... Language *Learning and Technology, 9*(1), 80-96.

Robertson, S., Wharton, C., Ashworth, C., & Franske, M. (1996). Dual device user interface design: PDA and interactive television. *Proceedings of CHI 1996, Vancouver* (pp. 79–86).

Sharples, M. (2000). The design of personal mobile technologies for lifelong learning. *Computers and Education, 34,* 177-193.

Sherington, R. (1973). *Television and language skills.* UK: Oxford University Press.

Tan, T., & Liu, T. (2004). The mobile interactive learning environment (MOBILE) and a case study for assisting elementary school English learning. *The 4th IEEE International Conference on Advanced Learning Technologies ICALT 2004* (pp. 530-534).

Underwood, J. (2002). Language learning and interactive television. In *Proceedings of workshop on Future TV*, ITS 2002, San Sebastian.

Endnotes

[1]   Previous versions of some of the material in this chapter have been presented at mLearn 2005 (Cape Town) and EuroITV 2006 (Athens).

[2]   Although Meinhof (1998) and Broady (1997) are exceptions.

[3]   This is a similar to that set out by Purushotma (2005), who suggests a range of activities, from playing a video game to listening to a song, in which language learning is a natural side effect of activities whose main goal is media use and consumption.

Chapter III

A Hybrid Strategy to Personalize the Digital Television by Semantic Inference

Yolanda Blanco-Fernández, E.T.S.E. Telecommunicación, Vigo, Spain

Jose J. Pazos-Arias, E.T.S.E. Telecommunicación, Vigo, Spain

Alberto Gil-Solla, E.T.S.E. Telecommunicación, Vigo, Spain

Manuel Ramos-Cabrer, E.T.S.E. Telecommunicación, Vigo, Spain

Martín López-Nores, E.T.S.E. Telecommunicación, Vigo, Spain

Abstract

The digital TV (DTV) will bring a significant increase in the number of channels and programs available to end users, with many more difficulties for them to find interesting programs among a myriad of irrelevant contents. So, automatic content recommenders should receive special attention in the following years to improve the assistance to users. However, current techniques of content recommenders have important well-known deficiencies, which complicates their wide acceptance. In this paper, a new hybrid approach for automatic TV content recommendation is proposed based on the so-called Semantic Web technologies, that significantly reduces those deficiencies. The strategy uses ontology data structures as a formal representa-

tion both for contents and users' profiles. The approach has been implemented in the AdVAnced Telematics search of Audiovisual contents by semantic Reasoning (AVATAR) tool, a new TV recommender system that makes extensive use of well-known standards, such as TV-Anytime and Web ontology language (OWL). Also, an illustrative example of the kind of reasoning carried out by AVATAR is included, as well as an experimental evaluation of the performance achieved.

Introduction

The denomination of DTV has been traditionally linked to greater quality of audio and video, which led many people to confuse DTV technology with high definition television (HDTV). But DTV has a unique strength that lies within its capability to broadcast data and telematics applications along with the audiovisual contents. Those applications, running on the users' receivers, are envisaged to cause a revolution in the very conception of the television.

To enable the execution of telematics applications, the new DTV receivers will be endowed with interactive and computational capabilities that will turn them into vehicles to access the information society. This way, it will be possible to fight the worrisome digital divide that is starting to show up in the developed countries, due to the limited penetration of the Internet in homes: nowadays, the information society is mostly accessed through Internet-enabled personal computers, and, as proved by data from InternetWorld Stats (http://web.archive.org), the penetration figures in homes seem to be reaching their peak (around 35% in Europe and 67% in the U.S.) as the growing rate is slowing down (from 29% in 2001 to 19% in 2004 in Europe, and from 14% to 7% in the U.S.). Knowing this, the fact that television is present in nearly every household in developed countries—being a familiar device for everyone—leads to thinking of DTV as the most likely means to overcome the commented barrier.

Additionally, after the establishment of the digital video broadcasting (DVB) solutions as the broadcasting standards, it is foreseeable that the normalization in the DTV field will focus on the telematics applications to overcome the current incompatibility problems between software and receivers from different providers. Thus, the market should progressively evolve into a truly horizontal model with well-defined roles (content providers, service providers, digital platforms, network operators, receiver manufacturers, and users) contributing to reducing costs and attaining greater acceptance of these technologies.

In this new scenario, the users will have access from their homes to a great number of channels and services from different providers. Resembling what happened with the growth of the Internet, this huge number of channels will cause the users to be disoriented: Even though they may be aware of the potential of the system, they lack the tools to exploit it, not managing to know what contents and applications are available and how to find them. Popular and successful search engines arose on the Internet from the syntactic matching processes of the mid-1990s to the more recent approaches to the Semantic Web, which are nowadays the subject of intensive research (Antoniou & Van Harmelen, 2004).

Specifically, taking advantage from the experience gained in the Semantic Web, this paper presents AVATAR, a TV recommender system whose main novelty is a process of reasoning carried out on the TV content semantics (provided by means of metadata) and user preferences (defined in personal profiles). This information is formally represented by a Semantic Web methodology widely used to conceptualize a specific application domain and to infer knowledge from it. This way, in AVATAR we resort to an ontology about the TV domain implemented by means of the World Wide Web Consortium (W3C) OWL (McGuinness, 2004). This knowledge about the TV domain is incorporated in the proposed hybrid recommendation strategy. This technique is based on mixing some well-known personalization methods with novel semantic inference capabilities applied on our TV ontology.

The rest of the paper is organized as follows: Section 2 presents an overview of the techniques being used by the current recommenders of audiovisual contents. The TV ontology and the user profiles are described in Section 3 and Section 4, respectively. Section 5 explains the algorithmic details of the proposed recommendation strategy. Section 6 shows a sample use of AVATAR. Finally, Section 7 includes the main conclusions from this work and motivates directions of future work.

Current Approaches to Personalization

In the last two decades, recommender systems have heightened a growing interest in the research community. This is supported by a great diversity in the approaches that can be found as much in the TV domain (Ardissono, Kobsa, & Maybury, 2004) as outside (Adomavicius & Tuzhilin, 2005). However, two personalization strategies can be highlighted because of their outstanding popularity: content-based methods and collaborative filtering. The former recommended to target users' contents similar to those ones they liked in the past, whereas the latter suggested to each user those contents appealing to spectators with similar preferences.

Content-based methods require a metric to quantify the similarity between the users' profiles and the candidate programs for recommendation. To define such metric, appropriate descriptions of the considered contents must be available, which is usually a complex and time-consuming task. By its own nature, this kind of method suggests contents too similar to those ones known by the user, which leads to a limited diversity in the elaborated recommendations. This problem is especially serious regarding new users of the system, as their suggestions are based on immature profiles with a limited set of programs.

To reduce these limitations without loosing the advantages of content-based methods, it is usual to combine this technique with collaborative filtering. By its very own, collaborative filtering provides much more diverse recommendations, as it is based on the experience of users with similar preferences (usually named neighbors). Also, these methods do not require the aforementioned resource-demanding content descriptions, as they search correlations among the ratings the users assign to contents.

However, some limitations can also be identified in the collaborative filtering techniques. Firstly, this technique requires that some users have rated a specific content for it to be recommended. Because of this, some significant latency is observed from when a new content arrives in the system until it is suggested to some user, as it is necessary that a significant number of users have previously rated the new content. Second, the lack of flexibility to estimate the similarity among users (usually based on direct overlaps among the programs in their personal profiles) leads to the so-called sparsity problem. In this case, as the number of contents in the system increases, the probability that two users have watched the same content gets lower. This reduced overlapping among the users' profiles greatly hampers the discovery of similar users regarding their preferences, a critical step in collaborative approaches.

With the aim to combine the advantages of each technique and minimize their drawbacks, a significant number of examples can be found in the literature that use hybrid approaches (Burke, 2002). This is the case of this paper that defines a new personalization strategy by combining a content-based algorithm with a collaborative filtering one, improving their accuracy thanks to the addition of semantic inference capabilities. This strength permits us to define a flexible metric to assess the similarity between two TV contents, whose value depends on the number of semantic relationships discovered between them.

An Ontology about the TV Domain

Ontologies are widely used as conceptualizations of a given application domain, where the characteristic vocabulary is identified by means of concepts and relations among them. These kind of methodologies constitute an appropriate formal framework to reason and infer new semantic relations among the concepts represented there. This attribute, along with the possibility to share and reuse the formalized knowledge in distributed environments, makes the ontologies a key technology in the Semantic Web.

In the specific case of the AVATAR tool, we have implemented an ontology describing the TV domain by means of OWL. The core of this ontology is a class hierarchy that identifies different kinds of TV programs, being the *TV Contents* class, the more general one, on the top of the hierarchy. From it, more specific categories are defined as successive descendants until reaching the more concrete categories, known as leaf classes, situated in the lowest level of the hierarchy. Note that we impose a tree-like structuring in our TV contents hierarchy, so that each class has only one superclass in it. In Figure 1, a reduced subset of the aforementioned hierarchy can be observed, together with the existing IS-A relationships among the classes.

All the classes shown in Figure 1 denote general content categories. Specific TV programs belonging to a given category correspond with specific instances of such classes. In fact, each program can be classified into several classes belonging to the above hierarchy. Regarding these instances and their semantics, it is worth noting that AVATAR uses the TV-Anytime specification ("TV-Anytime," 2001), which describes metadata attributes for audiovisual contents such as cast, topic, time, geographical information, and so forth. Such attributes, hereafter named *semantic characteristics,* are also instances in our ontology. Contrary to hierarchies, that only contain IS-A relations between the concepts, ontologies permit us to also define other relations between classes and instances by means of properties. This way, each program in the ontology will be related to its respective semantic characteristics through explicit properties (i.e., *hasActor, hasActress, hasTopic, hasTime, hasPlace,* etc.). These properties permit AVATAR to infer hidden knowledge in the ontology, to be applied in the proposed strategy.

Figure 1. Excerpt from our TV content hierarchy

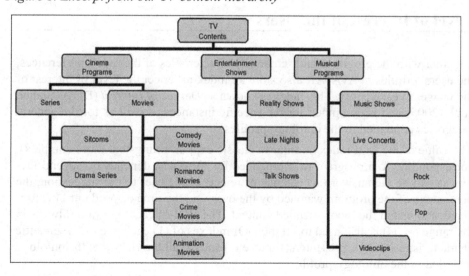

The Users' Profiles

As AVATAR reasoning involves the semantics of contents as well as the users' preferences, a formal representation of the latter is also needed to apply the inferential processes to them. Such preferences contain the programs that the user liked or disliked (positive and negative preferences, respectively) as well as those semantic characteristics relevant in a personalization environment (cast, genres, etc.). As this kind of information is already formalized in the TV ontology, our approach reuses such knowledge to model the users' preferences (for that reason they are stored in the so-called *ontology-profiles*).

Ontologies have already been used as tools to model users in previous works (Middleton, 2003). However, the distinctiveness of our ontology-profiles lies within their dynamic nature: they are progressively built as the system obtains additional knowledge about the user's preferences.

The Construction of the User's Ontology-Profile

This progressive building of the user's ontology-profile is carried out by adding to the profile only those classes (and their upper class hierarchy) that identify the specific contents associated to the positive and negative preferences of the user. When AVATAR knows a new content related to the user U, it adds to his/her profile (P_U) several items: (1) the instance that identifies the content in the ontology, (2) the class of that instance and its ancestors until *TV Contents* (Figure 1) and, (3) its main semantic characteristics.

Level of Interest of the Users

Together with the programs and semantic characteristics of the user's preferences, the users' profiles in AVATAR also store an index to reflect the level of interest of the viewer in each entity. This index is known as *Degree of Interest* (DOI) (Blanco et al., 2005) and is computed for the specific instances as well as for the general categories contained in the ontology-profiles.

The value of the DOI index (similar to classic explicit ratings but more complex), corresponding to a program recommended by AVATAR, depends on several factors as can be the answer (acceptance or reject) of the user to the suggestion, the percentage of the program watched by the user, and the time elapsed until the user decides to watch the recommended content. The DOI of each program (always in the range [-1,1]) is also used to set the DOI indexes of (1) each one of its semantic characteristics (actors, presenters, topic, etc.) and (2) all the classes of the ontology included in the ontology-profile.

Regarding the computation of the DOI indexes corresponding to the classes included in the profile, our approach firstly computes the DOI of each leaf class and then it propagates these values through the hierarchy until reaching the *TV Contents* class. The DOI of each leaf class is computed as the average value of the DOI indexes assigned to the programs, included in the profile, that belong to that class. To propagate those values through the ontology hierarchy, we adopt the approach proposed in Ziegler, Schmidt-Thieme, and Lausen (2004), that leads to equation (1):

$$ DOI\ (C_m) = \frac{DOI\ (C_{m+1})}{1 + \#\ sib(C_{m+1})} \tag{1} $$

where C_m is the superclass of C_{m+1} and $\#sib(C_{m+1})$ represents the number of siblings of the class C_{m+1} in the hierarchy of TV contents.

As a result of equation (1), this approach leads to DOI indexes higher for the superclasses closer to the leaf class whose value is being propagated, and lower for that class closer to the root of the hierarchy (*TV Contents* class). In addition, the higher the DOI of a given class and the lower the number of its siblings, the higher the DOI index propagated to its superclass.

A Hybrid Recommendation Strategy

Our strategy combines the content-based methods and the collaborative filtering, enhancing them with semantic inference capabilities. Its main goal is to decide to which users (named active users) a particular program must be recommended (named target content). By virtue of its hybrid nature we can identify two phases in our approach. Firstly, a content-based phase is applied, in which the approach assesses if the target content is appropriate for each active user by considering their personal preferences. In this case, the program is suggested to these users, whereas the remaining viewers are evaluated in a second stage based on collaborative filtering.

Content-Based Phase

Given an active user U and a target content a, this phase quantifies a level of semantic matching between this content and his/her preferences defined in the ontology-profile P_U (represented as match (a, U)). The more similar program a is to those contents most appealing to the user U, the greater the obtained semantic matching value. In order to measure this resemblance, we propose a flexible metrics, named *semantic similarity,* included in equation (2):

$$\text{match}(a,U) = \frac{1}{\#N_U} \sum_{i=1}^{\#N_U} \text{SemSim}(a,c_i) \cdot \text{DOI}(c_i) \tag{2}$$

where c_i is the i-th content defined in the profile P_U, DOI (c_i) is the level of interest of U regarding c_i, and $\#N_U$ is the total number of programs included in P_U.

Traditional approaches just use the hierarchical structure to quantify the semantic similarity between two concepts from a taxonomy, that is, they are only based on explicit IS-A relations established in the hierarchy (Ganesan, Garcia, & Widom, 2003; Lin, 1998; Resnik, 1999). These approaches hamper the kind of complex reasoning our intelligent system requires, and for that reason we redefine this traditional semantic similarity metrics.

As far as we know, our approach is the only one that combines the IS-A relationships with the inference of more intricate ones, discovered from the properties defined in our TV ontology. Thus, to compute the semantic similarity between the target content a and a given program b, our proposal considers both the explicit knowledge represented in the TV ontology and the implicit knowledge inferred from it. The proposed semantic similarity is composed of two components based on these two kinds of knowledge: hierarchical and inferential similarity, combined by means of a factor $\alpha \in [0, 1]$, as shown in equation (3).

$$\text{SemSim}(a,b) = \alpha \cdot \text{SemSim}_{\text{Inf}}(a,b) + (1-\alpha) \cdot \text{SemSim}_{\text{Hie}}(a,b) \tag{3}$$

The Hierarchical Semantic Similarity

The value of the hierarchical similarity between two programs depends only on the position of the classes they belong to in the content hierarchy. In order to define its analytical expression, we use two concepts from the graph theory: depth and lowest common ancestor (LCA).

* The depth of an instance (that identifies a specific program) is equal to the number of IS-A relations between the root node of the hierarchy (TV Contents) and the class the instance belongs to.
* On the other hand, being a and b two programs, the LCA between them (represented as $\text{LCA}_{a,b}$) is defined as the deepest class that is ancestor of both the class of a and the class of b.

Semantic similarity is defined by equation (4).

$$\text{SemSim}_{\text{Hie}}(a,b) = \frac{\text{depth}(\text{LCA}_{a,b})}{\max\ (\text{depth}(a), \text{depth}(b))} \qquad (4)$$

- SemSim$_{\text{Hie}}$ (a,b) is zero if the LCA between the two programs is TV Contents (which has null depth). Otherwise, the more specific the LCA$_{a,b}$ is (i.e., deeper), the greater the similarity value.

- The closer the LCA$_{a,b}$ is to both contents' classes in the hierarchy, the higher the SemSim$_{\text{Hie}}$ (a,b) is, because the relation between a and b is more significant.

The Inferential Semantic Similarity

The inferential similarity is based on discovering implicit relations between the matched programs. These relations are inferred between those TV contents that share semantic characteristics (e.g., cast, time, geographical information, topics, etc.). Thus, we consider that two programs a and b are related when they are associated by properties to instances—equal or different—of a same leaf class. In case these instances are equal, we say that the programs are related by a union instance; otherwise they are associated by a union class.

- The union instances can identify any of the semantic characteristics of the compared programs. For example, two movies starring the same actor are related by the union instance that identifies him in the ontology.

- On the contrary, the instances of the union class can only identify some of these characteristics. Specifically, those ones to which a higher flexibility is allowed in the comparison (e.g., topics or temporal and geographical locations). So, our approach can implicitly relate a World War I movie and a documentary about World War II. This association appears because both contents are related by means of the property *hasTopic* to two different instances of the union leaf class *War Topics* (*World War I* and *World War II* instances).

We define equation (5) to compute the value of SemSim$_{\text{Inf}}$ (a,b):

$$\text{SemSim}_{\text{Inf}}(a,b) = \frac{1}{\#\text{CI}_{\text{MAX}}(a,b)} \sum_{k=1}^{\#\text{CI}(a,b)} \text{DOI}(i_k) \qquad (5)$$

where #CI(a,b) is the number of common instances between a and b (union instances and instances of a union class), i_k is the k-th of them, and #CI$_{\text{MAX}}$ (a,b) is the maximum number of possible common instances between a and b (i.e., the minimum between the number of their semantic characteristics).

According to equation (5), the higher the number of union instances and union classes between both contents, and the greater the DOI of these common instances in the user's profile we are comparing to a, the higher the value of $\text{SemSim}_{\text{Inf}}$ (a,b).

When the matching levels for all of the active users have been computed by equation (2), AVATAR suggests the target content to each user with a level over a given threshold β_{Match}. The remaining users are candidate for the collaborative phase.

Collaborative Filtering Phase

In this phase, our approach predicts the level of the interest for each candidate user U_c with respect to the target content a (represented as Pred [U_c,a]), based on his/her neighbors' preferences. To this aim, the existing collaborative approaches only consider those neighbors who have rated the target content.

Our strategy differs from these methods because it considers the full neighborhood of each candidate user, during the semantic prediction process. So, if a neighbor knows the program a, our collaborative phase uses the specific DOI he/she defined in his/her profile. Otherwise, his/her level of interest is predicted from the semantic similarity between his/her preferences and this content a.

In order to create the neighborhood of each candidate user, our approach uses the so-called rating vectors of users, whose components are the DOI indexes in the user's profile for the classes in the hierarchy of contents. For those classes of the hierarchy not defined in the profile, we use a zero value.

After the rating vectors of each user are computed, our approach uses the Pearson-r correlation (Middleton, 2003) in order to compare them, as shown in equation (6), where $\overline{v_j}$ and $\overline{v_k}$ are the mean values of the v_j and v_k rating vectors, extracted from P_j and P_k, respectively.

$$\text{corr}(P_j, P_k) = \frac{\sum_r (v_j[r] - \overline{v_j}) \cdot (v_k[r] - \overline{v_k})}{\sqrt{\sum_r (v_j[r] - \overline{v_j})^2 \cdot \sum_r (v_k[r] - \overline{v_k})^2}} \qquad (6)$$

Notice that rating vectors do not require that two users have watched the same programs to detect if they have similar preferences. It is only necessary that the programs included in their profiles belong to the same classes in the content hierarchy. This represents a substantial reduction of the sparsity problem, typical in collaborative systems, as we said in Section 2.

Finally, the M viewers with higher correlation value with respect to the considered user form his/her neighborhood. Once all of the candidate user's neighborhood has been formed, our collaborative phase computes the semantic prediction value for each one of them by applying equation (7):

$$\text{Pred } (a, U_c) = \frac{1}{M} \sum_{k=1}^{M} \delta (\mathcal{N}_k) \cdot \text{corr } (U_c, \mathcal{N}_k) \tag{7}$$

where M is the neighborhood size; corr (U_c, N_k) is the Pearson-r correlation between the ratings vectors of the candidate user U_c and his/her k-th neighbor $\mathcal{N}_k$; and $\delta (\mathcal{N}_k)$ is a factor whose value depends on whether this neighbor has watched the program a. In case he/she does, $\delta (\mathcal{N}_k)$ is the DOI of a in $\mathcal{N}_k$'s profile; otherwise, it is the level match (a, $\mathcal{N}_k$) computed in the content-based phase.

According to equation (7), the value predicted to recommend a to U_c is greater when this content is appealing to his/her neighbors, and when their respective preferences are strongly correlated. Finally, note that we also define a threshold β_{Pred} (in [0,1]) for deciding if a is recommended to each candidate user.

An Example

In this section we show an example of the AVATAR reasoning process extracted from the experimental evaluations presented in Section 7. Due to space limitations, some simplifications are assumed: We have reduced both the number of semantic characteristics of each TV content and the size of the user's neighborhood employed in the collaborative phase (we limit to M=2). In spite of these simplifications, the shown example highlights the utility of semantic inference to compare the user's preferences and the advantages of our method for the neighborhood formation.

In this scenario, we assume that the target content is *Dancing With the Stars*, a reality show in which several celebrities (the singer Robbie Williams and the actress Jennifer Lopez, among others) are coupled with professional dancers and evaluated by a judging panel of dancing experts, chaired by Chayanne. Additionally, the users involved in this example are U and his/her neighbors N_1 and N_2. The TV programs contained in their ontology-profiles, as well as their respective DOI indexes, are shown in Table 1. These contents are represented together with some semantic characteristics in Figure 2. So, for instance, we can identify the sitcom *Friends* involving the actors Jennifer Aniston and Matthew Perry, or the movie *The Whole Nine Yards* co-starred by this same actor and Bruce Willis.

Table 1. Some TV contents defined in users' profiles

User U	User N₁	User N₂
Rolling Stones concert (0.8)	*Music!* (0.9)	***Dancing With the Stars (0.6)***
Friends (1)	*Dance With Me* (0.85)	MTV Videoclips (0.8)
The Whole Nine Yards (0.85)	*Desperate Housewives* (0.65)	Monster-in-Law (0.85)
		Frasier (1)

As shown in Table 1, our approach detects that the nearest neighbors of U are N_1 and N_2, although none of them has watched exactly the same programs than U. Indeed, the Pearson-r correlations between U, N_1, and N_2 are high because the three viewers are interested in instances of several subclasses related to Musical Programs (Rock Live Concerts, Music Shows, and Videoclips, respectively, in Figure 1). In addition, the sitcoms are also appealing programs to all of them (*Friends, Desperate Housewives,* and *Frasier* in Table 1).

Let us start describing the implicit semantic relationships inferred by AVATAR from the user's preferences shown in Table 1 (represented by numbered arrows in

Figure 2. Subset of instances and semantic relationships inferred from our TV ontology

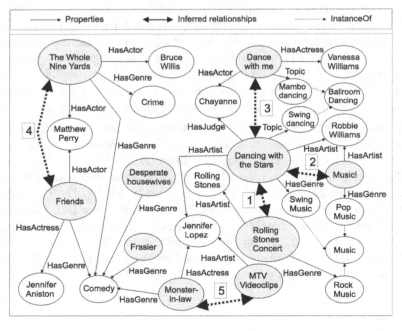

Figure 2). Next, we will see how the discovered knowledge is used in the recommendation process.

- Although the TV-Anytime metadata define different musical genres for *Dancing With the Stars* and Rolling Stones concert (Swing Music and Rock Music, respectively), our approach relates them by the *union class Music* (relationship 1 in Figure 2).

- The considered reality show and *Music!* are semantically related because both contents share the musical genre and involve the singer Robbie Williams (relationship 2 in Figure 2).

- The implicit relationship between *Dancing With the Stars* and the movie *Dance With Me* (relationship 3 in Figure 2) is due to the fact that these contents involve ballroom dancing and, besides, Chayanne participates in both of them. Obviously, approaches based on simple syntactic comparisons between the attributes of the considered programs would not uncover this kind of relationship.

The content-based phase. Our system compares U's preferences (Table 1) against the reality show Dancing With the Stars. According to Figure 1, the hierarchical similarity between these programs is zero because of the non-existence of an LCA between their respective classes other than TV Contents. On the other hand, the inferential similarity is not significant because AVATAR infers only one relationship between the Rolling Stones concert watched by U and the reality show (relationship 1). Note that this relation is established by means of an only instance of the union class Music. Because of this, equation (5) quantifies a very low inferential similarity between the U's preferences and the reality show. The combined value of both components is not high enough to exceed the threshold β_{Match}. Therefore, Dancing With the Stars is not offered to U, and, so, this user is a candidate for the collaborative phase.

The collaborative phase in AVATAR. Opposite to what happens with conventional approaches, our strategy considers all the neighbors of U in the collaborative phase, both those who have watched the reality show and those who do not know this content. In our scenario, as N_1 has not watched *Dancing With the Stars*, AVATAR predicts a specific level of interest for him/her by computing the semantic similarity between his/her preferences and the reality show. The aforementioned relationships 2 and 3 are used to compute this value. Before describing this process, note that the respective semantic characteristics of the contents *Music!* and *Dance With Me* (see Figure 2) are of interest for N_1, because those programs are very appealing to him/her. Specifically, considering the *union instances* and classes existing between both contents (Robbie Williams, Chayanne, and ballroom dancing in Figure 2) and their high level of interest in N_1's profile, equation (5) computes an inferential similarity greater than β_{Match}. Thus, the system suggests *Dancing With the Stars* to N_1.

Opposite to what happens in the existing collaborative approaches—where only one neighbor of U was moderately interested in *Dancing With the Stars*—our strategy predicts that this content is also appealing to N_1. As a consequence, the reality show is finally suggested to U. This prediction value, computed by equation (7), is high enough to exceed the threshold β_{Pred}, because all of U's neighbors are interested in the reality show, and also their preferences are strongly correlated to this user's ones.

Lastly, note that the existing collaborative approaches would not offer the reality show to U until the neighbor N_1 had watched and liked this program. Considering that *Dancing With the Stars* was suggested to this user thanks to our novel inference capabilities, it is clear that the existing proposals would never recommend this show to N_1. Thus, the reality show would also not be recommended to U, despite the fact that our approach discovers that it is appealing to both viewers.

Experimental Evaluation

To carry out the experimental evaluation of our hybrid approach, we have resorted to a group of 400 undergraduate students who rated a set of 400 TV contents with a specific DOI index in the range [-1,1]. These contents were shown together with a brief description, so that all users could judge the offered programs, even if they had never watched them.

Test Algorithms

Our goals were: (1) to evaluate the accuracy of the recommendations offered by our hybrid strategy, and (2) to compare these results with those achieved by a variant of this strategy, that only differs in the similarity metric employed. In fact, it is also a two-phase proposal, whose metric is based on detecting the so-called *association rules* between the matched TV contents, hence increasing their similarity value.

As shown in O'Sullivan, Wilson, and Smyth (2004), these rules are automatically extracted by the a priori algorithm (Agrawal & Srikant, 1994), which is applied on a training set of profiles containing several programs. The aim is to discover common patterns between these contents that yield the association rules.

For example, being A and B two TV contents, a rule of the form $A \Rightarrow B$ means that if a profile contains the program A, it is likely to contain the program B as well. In fact, the probability that this rule holds, named *rule confidence*, is the percentage of profiles containing B given that A occurs. This way, as in O'Sullivan et al. (2004), the similarity between A and B is the confidence of the rule that involves both contents, as long as this value is greater than a threshold confidence (set in a priori).

Test Data

Because of the presence of the association rules in our experiments, part of the user's preferences must be used as a training set for the a priori algorithm. So, it will be possible to discover relations between the 400 contents considered in this evaluation and to use them to measure their similarity. For that reason, the initial 400 users were divided into two groups:

- *Training users* (40%). The training profiles used in a priori contain exactly the programs these users have rated with positive DOI indexes at first. Regarding the threshold confidence, we assumed the value 0.1 used in O'Sullivan et al. (2004).

- *Test users* (60%). These users chose 20 contents out of the initial 400 programs (the 10 contents they found most appealing, and the 10 that these users most disliked). In this process, we saw a low overlapping between the programs selected by the users, thus causing a significant sparsity level. On one hand, these 20 contents were used to construct the profiles on which the two evaluated approaches were applied. On the other hand, the remaining 380 programs (named *evaluation data*) were hidden and used to measure the recommendation accuracy. Comparing these contents to those offered by each considered approach, we achieved an accuracy metric appropriate for our evaluations.

Methodology and Accuracy Metrics

Once the experimental data have been presented, we describe the procedure applied on the test users.

- Firstly, we applied each approach by using as target contents the 400 contents available in the initial set, as we explained in Section 5 (the neighborhood size M=30 and the thresholds $\beta_{Match} = \beta_{Pred} = 0.6$ are used).
- Next, the *recall* for each user was measured, computed as the ratio of programs rated with positive level of interest in the test user's evaluation data that were suggested by the corresponding approach. We chose this metric because recall is a stringent accuracy criterion. In this regard, our main goal is to check whether the semantic relations inferred in our strategy are able to discover all

the programs appealing to the user. A precision metric is not used in this evaluation as it would refer to the proportion of correct-to-all recommendations thus not being as revealing an accuracy measure, according to the aforementioned aim. This same decision was taken in O'Sullivan et al. (2004).

- Finally, the average recall values (computed over all of the test users) for the two approaches were represented, as shown in Figure 3. Note that we have evaluated different values for the parameter α in equation (3).

Discussion on the Experimental Results

The results from Figure 3 indicate that the semantic inference increases greatly the recommendation accuracy. This way, in spite of huge sparsity level in the used data (89%), we have obtained a 26% difference between the *average recall* of both approaches, computed by considering the 240 test users (58% of average recall for $\alpha = 0.7$ in our proposal, against 32% in rule-based approach).

This significant increase is due to the fact that the semantic inference managed to uncover some relationships between TV contents that went unnoticed for the rule-based approach.

This is because our implicit relationships are inferred from the complete and precise knowledge represented in the OWL ontology, whereas the association rules depend on the patterns that a priori finds in the training profiles.

Figure 3. Recall values for the evaluated approaches

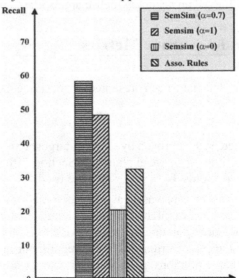

On the other hand, regarding the α parameter, the best results are obtained when the two components of the semantic similarity are considered (see equation (3)), and when the inferential one weighs more, as shown in Figure 3 with $\alpha = 0.7$. Specifically, the recall values for $\alpha = 0$ evidence again the aforementioned benefits of the semantic inference. In this scenario—where we omit the inference of new knowledge and only consider the explicit IS-A relationships in Figure 1—we see that the suggestions are much less accurate than in the cases where $\alpha \neq 0$. Also, only for $\alpha = 0$ the recall values obtained by the rule-based approach are greater than those achieved by the semantic inference.

Conclusion and Future Work

In this paper, we have proposed a recommendation strategy for a TV intelligent assistant, based on mixing two well-known personalization approaches: content-based methods and collaborative filtering. Our approach complements those personalization techniques with a novel process of inference from the user's preferences and the TV content semantics, formally represented in an OWL ontology about the TV domain. The results extracted from our empirical evaluations show significant increases in the recommendation accuracy, in spite of a very high sparsity level in the used data. Specifically, we have achieved a 26% difference between the average recall of the two considered approaches. This evaluation proves that the inferential processes carried out in AVATAR are able to discover appealing relationships between the TV contents, never uncovered in approaches lacking in semantic reasoning capabilities.

Also, the proposed hybrid strategy alleviates some well-known drawbacks related to both the collaborative and the content-based approaches. Regarding the former ones, our proposal reduces greatly the sparsity problem, thanks to a technique that uses a hierarchy to identify the general categories which the user's preferences belong to, instead of identifying specific programs. So, we can exploit this hierarchical structure in order to generate overlaps between profiles containing different TV contents. Our proposal also alleviates the latency problem of the collaborative systems, since whenever a new content arrives to the system this can be suggested to the users with no delay.

The lack of diversity associated to the content-based filtering is also greatly reduced in our proposal. For that purpose, we have used a flexible metric to compare TV contents, named semantic similarity. This metric is based on inferring semantic relationships between the matched programs; so, the greater the number of inferred relationships the higher the measured similarity. To the best of our knowledge, no existing approach considers in this kind of measure both the explicit IS-A relations defined in a TV content hierarchy and the inference of more complex relationships.

The latter are discovered between programs that share semantic characteristics relevant for the users (e.g., cast, time, geographical information, topics, etc.).

This approach is flexible enough to be used both in a single-user model of recommendation and in a scenario where the target audience is a group of people. In the last case, our technique selects target contents recommended to all the users, thus fulfilling the group preferences. Additionally, note that our approach is general enough to be extended to other personalization applications, outside the TV domain.

Finally, we plan to carry out new experiments in order to compare our strategy to other existing approaches, such as the traditional user-based collaborative filtering, and the item-based collaborative filtering proposed in Sarwar, Karypis, Konstan, and Riedl (2001).

Acknowledgment

This work has been supported by the EUREKA ITEA Project PASSEPARTOUT and by the Spanish Ministry of Education and Science Project TSI2004-03677.

References

Adomavicius, G., & Tuzhilin, A. (2005). Towards the next generation of recommender systems: A survey of the state-of-the-art and possible extensions. *IEEE Transactions on Knowledge and Data Engineering, 17*(6), 739-749.

Agrawal, R., & Srikant, R. (1994). Fast algorithms for mining association rules. In *Proceedings of 20th International Conference on Very Large Data Base* (pp. 487-499). Santiago de Chile.

Antoniou, G., & Van Harmelen, F. (Eds.). (2004). *A semantic Web primer*. Cambridge, MA: MIT Press.

Ardissono, L., Kobsa, A., & Maybury, M. (Eds.). (2004). *Personalized digital TV: Targeting programs to individual viewers*. Kluwer Academic Publishers.

Blanco, Y., Pazos, J., Gil, A., Ramos, M., López, M., & Barragáns, B. (2005). AVATAR: Modeling users by dynamic ontologies in a TV recommender system based on semantic reasoning. In *Proceedings of 3rd European Conference on Interactive TV*. Aalborg.

Burke, R. (2002). Hybrid recommender systems: Survey and experiments. *User Modeling and User-Adapted Interaction, 12*(4), 331-370.

Ganesan, P., Garcia, H., & Widom, J. (2003). Exploiting hierarchical domain structure to compute similarity. *ACM Transactions on Information Systems, 21*(1), 64-93.

Lin, D. (1998). An information-theoretic definition of similarity. In *Proceedings of 15th International Conference on Machine Learning* (pp. 296-304). Wisconsin.

Middleton, S. (2003). *Capturing knowledge of user preferences with recommender systems.* Unpublished doctoral dissertation, University of Southampton.

O'Sullivan, D., Wilson, D., & Smyth, B. (2002). Using collaborative filtering data in case-based recommendation. In *Proeedings of 15th International Artificial Intelligence Research Society Conference* (pp. 121-125). Florida.

Resnik, P. (1999). Semantic similarity in a taxonomy: An information-based measure and its application to problems of ambiguity in natural language. *Journal of Artificial Intelligence Research, 11*(1), 95-130.

Sarwar, B., Karypis, G., Konstan, J., & Riedl, J. (2001). Item-based collaborative filtering recommendation algorithms. In *Proceedings of 10th World Wide Web Conference* (pp. 285-295). Hong Kong.

TV-Anytime specification series: S-3 on metadata. (2001). Retrieved from http://www.tv-anytime.org

Ziegler, C., Schmidt-Thieme, L., & Lausen, G. (2004). Exploiting semantic product descriptions for recommender systems. In *Proceedings of the Semantic Web and Information Retrieval Workshop.* Sheffield, UK.

Chapter IV

An Approach for Delivering Personalized Advertisements in Interactive TV Customized to Both Users and Advertisers

Georgia K. Kastidou, University of Waterloo, Canada

Robin Cohen, University of Waterloo, Canada

Abstract

In this chapter, we present a model for delivering personalized ads to users while they are watching TV shows. Our approach is to model user preferences, based on characterizing not only the keywords of primary interest but also the relative weighting of those keywords. We combine the results of two separate agents: TV Monitoring Agent (TMA) tracks the kind of shows being watched by the user, for how long, and on what days; Internet Monitoring Agent (IMA) captures the keywords of interest to the user, based on browsing activity. The conclusions reached by these two agents are merged into one representation, compared to a characterization of possible ads to be delivered, and adjusted to fit into required time slots. We consider as well the case of providing ads for an entire household of users, making use of the collection of individual profiles. We discuss how our approach results not only benefit users but also the benefit to advertisers.

Introduction

This chapter focuses on the problem of providing personalized advertisements to people who are watching interactive TV (iTV) shows. The aim is to decrease the cost of TV services by increasing the amount of advertisements the users watch during TV shows. The problem is that users may become annoyed by the ads that are provided. We develop an approach that allows a channel provider to select ads from a predetermined set in such a way that the profiles of potential users in the household are well matched to each ad. We, in fact, allow for these choices to be made dynamically as the show is being broadcast.

In particular, we propose acquiring information about each user both from a monitoring of their viewing habits (not only which shows but for how long and on what days) and an analysis of their Internet browsing habits (initial Web pages and others linked to them). We use a keyword-based approach to represent these user models, incorporating methods for weighting the various keywords that indicate the user's interests. An especially important aspect of our model is representing a user's interest in the keywords of a show more heavily when a user actually watches a significant percentage of that show and discounting possible conclusions about the user's interests in a show, when only a small percentage of that show has been watched. The current show and all its potential advertisements are also represented in terms of keywords, enabling an effective selection of ads that are personalized to the users.

We then move on to discuss the important case of multiple viewers in a single household. Whereas, most efforts to personalize TV assume there is a single user to address, we allow for more than one person to be viewing a show at the same time. We propose an algorithm for selecting ads that is open to potential differences among the possible viewers and also clarify how our model can begin to account for cases where the user is in the household but in fact is not watching the TV at the time of the modeling.

In all, with personalization we provide a method for delivering ads that should be agreeable to users, and with an effective procedure for selecting ads that should please a set of potential viewers, we allow not only satisfied advertisers but satisfied channel providers, who may then operate at a lower overall cost.

Background

Personalization is the method that captures the users' profiles (preferences) in order to use them as a filter whenever a user searches for specific data.

For instance, assume that there are two teenagers, a Greek and a Canadian, who are interested in soccer. The problem is that in Greece, and in general in Europe, most of the time soccer is called football, while football in Canada, and in general in North America, is a totally different sport. Thus, if the Greek and the Canadian teenager use a search engine on the Internet and enter soccer and football respectively, the results in each case will contain a mix of two totally different sports; consequently, both of the users will have to go through much irrelevant and annoying information.

This problem can be solved with the usage of a personalized system. A personalization system in this case would be able to recognize that both of the teenagers refer to the same sport that in Greece is called football and in Canada soccer; thus, the system would bring only information relevant to this sport.

Personalization in iTV has been applied in many different domains. The most interesting are: (1) program recommendation (Ardissono et al., 2003; Kurapati, Gutta, Schaffer, Martino, & Zimmerman, 2002; Xu, Zhang, Lu, & Li, 2002), (2) news customization (Merialdo, Lee, Luparello, & Roudaire, 1999), and (3) advertisement customization (Lekakos & Giaglis, 2002). In the context of our topic, personalized ads for the iTV, the goal is to show ads that will interest the viewers.

Our research was motivated by other efforts in delivering personalized television to users and on inserting personalized ads into iTV. The work of Ardissono et al. (2003) is important because it combines both explicit user preferences and ones acquired implicitly by observing a user's TV viewing habits, in order to generate a ranked list of programs to recommend for the user.[1] For our research, we need to select advertisements during shows (so not simply making recommendations which a user can accept or reject). We elect to emphasize methods to implicitly acquire user preference information (dynamically), but expand our consideration to include Internet habits as an additional source of information. We also develop methods to carefully set the weights on the list of items representing the user's interests.

One valuable approach for personalizing ads to users is presented in Lekakos and Giaglis (2002) and Lekakos, Papakiriakopoulos, and Chorianopoulos (2001), where the authors employ a collaborative filtering approach to recommend ads to users, based on the preferences of other like-minded users. In this way, the ads being generated are really customized to an entire set of possible viewers. This suggestion provides some motivation for us to study entire households of viewers, but we want to allow for a more heterogeneous group of users and address how to merge different user profiles when selecting ads. We also seek to develop a more efficient architecture for delivering the ads, sending only those that will be broadcast to the users.

The suggestion of accounting for both past and current user preferences, when selecting ads for presentation, is discussed in Thawani, Gopalan, and Sridhar (2004). This model also employs a user profile in order to customize the ads selected to the preferences of the user. The particular algorithms, driving the acquisition of user's preferences, are not sufficiently detailed, nor is the method for matching user-program-advertisement similarities. And while the approach should be applicable to households of multiple users, there is no explicit discussion for how to distinguish the different users. We decided to explore in more detail, within our own research, how to address the challenge of multiple viewers within our ad selection procedure.

Important research in how to address the interests of groups of viewers of iTV has been conducted by Masthoff (2004). The emphasis of this work is to explore various strategies for addressing the competing interests of the group members that select content for viewing. Our proposal for multiple viewers considers a new concern: to what extent advertisers can extend the current interest of a group of viewers by offering ads that are targeted more to other individuals in their group.

The Problem

The problem we address in this work is the design of a system which will provide users with cheap TV services that have less annoying advertisements. The goal of our approach is to allow a number of customized advertisements to be broadcast. The customization of the ads should be done in a way that makes the selected ads interesting to the viewers, thus less annoying, but also satisfies advertisers' expectations. We believe that although advertisers wish their ads to be seen mainly by short-term potential buyers, they are also interested in broadcasting to people who change their interest to become potential buyers.

To achieve our goal, we propose a mechanism that: (1) considers viewers' profiles, based on their TV and Internet preferences for selecting the advertisements they watch, (2) maximizes the advertisements' effect by associating them with specific shows that might contain the advertised products, and (3) accepts that it might be difficult to recognize the people who are watching TV and turn this to an advantage (using possible "wrong" system predictions to expand viewers' profiles and allow ads on topics to be shown to these users).

The approach that we propose is based on the combination of the user's TV profile and the user's Internet profile. The inputs of our approach are: (1) the viewers' TV preferences, captured through monitoring viewers' TV habits, (2) viewers' Internet preferences, captured through monitoring their Internet behavior, (3) the potential advertisements for the specific commercial break, and (4) the time duration of the commercial break. We also assume that we are starting with a particular show that requires an ad.

The output is a set of ads that fits in the commercial time slot and are relevant to the viewers' profiles, to the current show's content, and to the advertiser's expectations of appealing to the viewers.

More specifically, we assume that one viewer is watching a show and a commercial break is going to be broadcast in the next few minutes. The system will ask the two agents that are responsible for keeping the viewer profile to send a part of their profile to a coordinator agent (the COD agent). This agent merges the two lists of keywords that reflect the viewer's interests and forwards them to the channel provider. As soon as the channel provider receives the list, it finds the advertisements that are most likely to be interesting to the viewer and fits them into the advertisement break time slot and broadcasts them to the viewer. In the case where more than one viewer is watching TV, the COD agent receives their preference lists and merges them into one, and then follows the same procedure as in the case of the one viewer.

One important assumption is that there are a limited number of persons in the house, and the persons who live in the house do not change without notifying the system. We also assume that the system knows the persons inside the house. The scope of these assumptions is to make sure that this is not a public TV where the viewers change frequently over time, but instead an environment like a family style house.

The Architecture of the System

An Overview

Before presenting our approach in detail, we make some conventions. We will use the word *show* for referring to anything the viewer can watch on TV except news and advertisements. Thus, a show can be a movie, a show, a TV game, and so forth.

Each show is associated with some keywords. A word can be characterized as a *keyword* if it can describe the content of show. For instance, for a theatrical musical show some of the keywords would be music, sing, dance, actors, theatre, and so on.

Each keyword is associated with a *weight* that reflects the importance of the keyword in the show's description. A weight can have any value inside the [0,1] interval. The larger a weight the more significant in the show's description the keyword is. For instance, in our musical show the weight might be (music, 1), (sing 1), (dance, 0.86), (actors, 0.6), (theatre 0.9), and so forth.

Each of the viewers has a profile that consists of a set of keywords that can be found either by explicitly asking the viewer or by monitoring the viewer's TV preferences, as well as by monitoring the viewer's Internet preferences[2] (which Web sites he/she visits more frequently, etc). Each of these keywords is associated with a weight according to the system's belief on how much it matches the viewer's profile.

System's Components

The system consists of the hardware and software components. The hardware components are the devices that are necessary for the system to work, and the software components are the software that needed to be installed and run on some of the devices for providing the functionalities of the proposed mechanism.

The software component of the system consists of three different categories of intelligent agents, the TMA, the IMA, and the COD. We will use the TMA, IMA, and COD notation for referring to those agents. In case more than one user lives in the house, each user has his/her own TMA and IMA.

The system hardware components are the TV, a device attached to the TV, which is called the set-top box, and a computer connected to the Internet.

The set-top box communicates through a local area network (LAN) with the computer.

Agents TMA and COD run on the set-top box, while agent IMA runs on the computer. As we have already mentioned, the COD agent collaborates with both of the IMA and TMA. The overall architecture is presented in Figure 1.

In the next three sections we discuss the functionalities of the software agents.

Figure 1. The architecture of the system

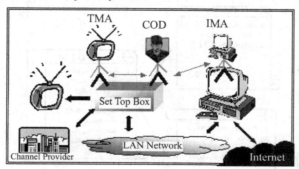

Agent TMA

TMA runs on the set-top box. A TMA monitors what the viewer watches on TV. It maintains a list of keywords that are associated with the shows the viewer has so far watched or is watching to reflect the user's preferences. For each keyword, the agent has a weight that reflects the system's belief on how much the keyword satisfies the viewer's preferences (Figure 2). Also, the TMA maintains—for each of the seven days of the week—the average time, $t_{DAT\_TMA}(i)$ (where i=1,...,7 [1=Sunday, 2=Monday, etc.]), the viewer watches TV.

TMA is responsible for changing the values of the weights in order to achieve the optimal distribution of the weights. As we will see in the *Updating TMA's Weights* section, the TMA monitors the viewer's behavior when he/she watches TV (i.e., how much time the viewer spent watching the show) and uses this information to update the weights' values, taking into account a number of factors.

Agent IMA

IMA is an agent that runs on the computer and monitors the viewer's preferences when he/she surfs the Internet. We assume that each viewer logs in to the computer using a personal login name and password. As in the case of TMA, IMA also maintains a list of keywords. These keywords are associated with the Web sites the viewer prefers to visit, and each of these keywords is associated with a weight that reflects the system's belief about how important the specific keywords are to the user, based on how often they visit Web sites with these keywords.

Figure 2. Merging TMA and IMA keyword lists

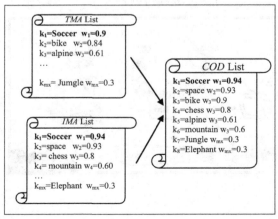

Finally, IMA also maintains—for each of the seven days of the week—the average time, $t_{DAT\_IMA}(i)$ where i=1,...,7 (1=Sunday, 2=Monday, etc.), the viewer usually surfs on the Internet.

COD Agent

COD agent is the agent that is responsible for sending to the channel provider the list of keywords that match the profile of the viewer who is watching the TV at that time.

More particularly, the channel provider sends a Viewers'_Profile_Request to the COD agent, requesting information about the viewers' preferences. Then, the steps the COD agent follows for providing the requested information are:

- **Step 1:** The COD agent sends a message to the viewer's TMA and IMA requesting:[3] (1) from the TMA, the m keywords (and their weights) with the highest weight values and the l keywords with the lowest weight values; and (2) from the IMA the r keywords (and their weights) with the highest weight values and the q keywords with the lowest weight values.
- **Step 2:** The TMA and IMA select the keywords the COD agent asked for, but before they send them, each of the TMA and IMA passes its list through a normalization function (we discuss this procedure in more detail in the *Normalization Function* section). The reason for normalization is that keywords in the TMA and IMA lists are to contribute equally for generating the merged list.
- **Step 3:** The COD agent receives the lists and merges them in one sorted list in descending order, according to the weights of their keywords. If a keyword is found in both of the lists, the COD agent chooses to keep and sort the specific keyword considering the weight with the highest value.[4]
- **Step 4:** Finally, the COD sends the merged list to the channel provider.

Selecting Advertisements for a Single Viewer

As we already mentioned, the channel provider maintains for each show in its database a set of keywords (and their weights) that describes the show. A short period of time before interrupting a show with a commercial break, the channel provider communicates with the corresponding COD agent, requesting the list of keywords and their weights from the viewer's profile. Then, the COD agent communicates with the viewer's TMA and IMA in order to receive the requested data and forwards them to the channel provider. The next step is for the channel provider to select the appropriate advertisements based on the COD agent's list, trying to match the advertisements to the viewer's profile and to the content of the show the viewer is currently watching.

Assume that L_{COD} is the list of the keywords with their weights that the service provider received from the COD agent, L_{SHOW} is the list that describes the current show's keywords with their weights, and P is the set of the potential advertisements for the specific time slot.

The channel provider creates a new list by merging the L_{COD} and L_{SHOW} into one list, called L_m, by taking the k_{COD} and k_{SHOW} keywords with the highest weights from the L_{COD} and L_{SHOW} lists, respectively, for a total of k_{MAX} keywords. The values of the k_{COD} and k_{SHOW} variables are given by the channel provider and their functionality is to determine the importance of the viewer's profile and show's keyword list in the selection of the advertisements.[5]

Crucial for the selection of the appropriate ads is the calculation for each advertisement $p \in P$ of a function called *reward*. The scope of the reward function is to find the advertisements that satisfy the viewer's and the show's keyword list the most. For each keyword in the L_m list it finds the summation of their weights in each of the possible advertisement's keyword list. For instance, considering the L_m list of Figure 4, the reward function for the advertisement e in Figure 3 is:

$$reward(e) = \text{Alpine keyword's weight}$$
$$+ \text{Mountain keyword's weight}$$
$$= .95 + .94 = 1.89$$

The reward function sums only the weights of keywords *Alpine* and *Mountain* because these are the only common keywords between the L_m list and the advertisement's keyword list. This is done for each of the possible advertisements (P) for this time slot of the show.

The next step is to select the advertisement that fits in the specific time slot. This is done by finding the advertisement that has the maximum reward. We then check if this advertisement can fit in the remaining time slot. If it does, we put it in the list of ads to be shown. Otherwise, we continue to search the possible ads to find the ad that has the next highest reward value. Note, that the ads will be broadcast in the order that they were selected.

The algorithm for selecting the appropriate ads is as follows:

Algorithm: ADVERTISEMENT_SELECTION

Input: L_{COD}, L_{SHOW}, P, $L_{ads}(p)$ $\forall p \in P$, $t_{ads}(p)$ $\forall p \in P$, t_{ad}, t_{min}, k_{COD}, k_{SHOW}.[6]

Output: A list S that contains the advertisements that will be shown.

- **Step 1:** Merge L_{COD} and L_{SHOW} in a list L_m by taking the k_{COD} keywords with the highest weights from the L_{COD} list and k_{SHOW} keywords with the highest weights from the L_{SHOW} list.

- **Step 2:** // Find the reward $\forall\, p \in P$

$$\forall\, p \in P$$

$$\forall\, k \in L_{ads}(p) \qquad\qquad // \text{ k: keyword}$$

$$\text{if } k \in L_m \text{ then} \qquad\qquad // \text{ w}_k\text{: weight of k in p}$$

$$\text{reward}(p) \leftarrow \text{reward}(p) + w_k$$

- **Step 3:** $Q \leftarrow P, t_{rm} \leftarrow t_{ad}, S \leftarrow P$

- **Step 4:** Find p^* where

$$p^* \leftarrow \underset{p \in Q}{\text{argmax}}(\text{reward}(p))$$

- **Step 5:** $Q \leftarrow Q - \{p^*\}$

$$\text{If } t_{ads}(p^*) \le t_{rm} \text{ then}$$

$$t_{rm} \leftarrow t_{rm} - t_{ads}(p^*)$$

$$S \leftarrow S \cup \{p^*\}$$

- **Step 6:** If $Q=\varnothing$ or $t_{rm} < t_{min}$ then

$$\text{return}(S, t_{rm})$$

$$\text{Else Go to Step 4}$$

For example, assume that John watches a movie that has to do with alpinists, and the keyword list of this particular movie is the following [(mountain, 0.99), (alpinist, 0.84), (bike, 0.6), (river, 0.4)].

Figure 3. Advertisements' keyword lists

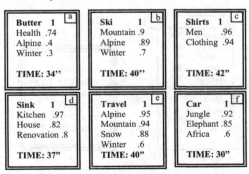

Now assume that the channel provider wants to interrupt the flow of the movie with a 110 second commercial break and that the potential advertisements for the specific time slot are those in Figure 3.[7]

The advertisement *a* advertises butter and the story takes place at a house in the mountains during winter, *b* advertises the ski sport, *c* advertises shirts, *d* advertises sinks, *e* advertises a travel office that organizes trips for alpinists, and finally *f* advertises a car, and the story of the advertisement takes place in a jungle in Africa.

The rewards for each of the potential ads are shown in Figure 4(b). According to the advertisements selection algorithm, the advertisements that will be broadcast to the viewer are: *e*, *b*, and *f*, with a total duration of 110 seconds. This means that John is going to watch the advertisements for the travel agency, the ski sport, and the car.

The Normalization Function

We use a normalization function in order to find how important each of the keywords in the viewer's preference list is in proportion to the rest of the keywords, in order for the merged list to better reflect the combination of the viewer's TV and Internet preferences.

In the TMA's case, the normalized value nw_i of weight w_i is:

$$nw_i = w_i / \sum_{k=1}^{tma\_max} w_k$$

where *tma_max* is the maximum number of keywords that the TMA can store.

Figure 4. (a) the L_m list and (b) the values of the reward function of each of the ads in Figure 3.

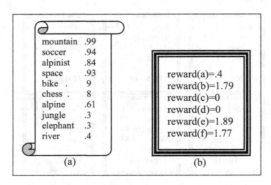

In IMA's case, the normalized value nw_i of weight w_i is:

$$nw_i = w_i / \sum_{k=1}^{ima\_max} w_k$$

where *ima_max* is the maximum number of keywords that the TMA can store.

Updating TMA's Weights

One very crucial part of our mechanism is the TMA weight update procedure. Before we present the update function we discuss some of the potential cases in watching a show, concerning the time a viewer starts to watch a show, the time he/she stops watching it, and how this reflects his/her interest in the show. More specifically, let us consider the time interval $[t_{ss}, t_{se}]$ in Figure 5(a), where t_{ss} is the time a show S started and t_{se} the time it ends. Consider now that a viewer X starts to watch this show. Three possible cases could be:

1. Viewer X starts watching the show some time during the time interval dt_{se} in Figure 5(b) and stops watching it sometime during the time interval dt_{ue}. The result is that in this case the viewer only watched a part of the show. The most likely cases are the following ones: (1) the viewer did not like the show and changed the channel, or (2) the viewer had to go so he/she shut the TV off or someone else changed the channel to watch after viewer X left, and (3) the viewer is sharing the TV with someone else and has to give up his/her turn. We could reach different conclusions about whether the viewer liked the show in each of those cases.

Figure 5. Viewer's start and stop cases

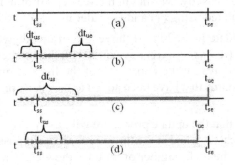

2.   Viewer X starts watching the show sometime during the time interval dt_{us} in Figure 5(c) and keeps watching it to the end. In this case it is more likely that viewer X liked the show. So, the keywords of the show must contribute significantly to his/her profile.

3.   Viewer X starts watching the show sometime during the time interval dt_{us} in Figure 5(d) and keeps watching it until t_{se}, just a few minutes before its end. In this case the viewer X: (1) may have liked the show but missed the end because he/she decided to change the channel and forgot to return to the show, or (2) might have left the room or fallen asleep while watching the previous show, so later just came back to turn off the current show. In the first case, the keywords of the show must contribute significantly to the viewer's profile while in the second case this makes less sense.

To provide viewers with advertisements that explicitly match their preferences it would be ideal to have a procedure that could distinguish what actually happened and update the weight accordingly, but this is difficult to achieve.

We can make some sensible adjustments. Our update function considers the time the viewer watched a show, the importance of each of the show's keywords, the percentage of the show the viewer watched, and the duration of the show in proportion to the average time the viewer watches TV during each specific day (Monday, Tuesday, etc.). So, each keyword i in the show's keyword list our update function[8] will be:

$$w_i \leftarrow w_i + sw_i * t_u^2 / (t_d * t_{DAT\_TMA}(z)) \tag{1}$$

where w_i is the weight of the keyword i in viewer's list, sw_i is the weight of the keyword i in the show's list, t_u is the time that the user watched the specific show, t_d is the time duration of the show, z is the current day the update procedure takes place ($z=1$ for Sunday, $z=2$ for Monday etc.), and finally $t_{DAT\_TMA}(z)$ is the average time the viewer watches TV during day z.

In case a keyword from the show was not in the viewer's profile it will be inserted. Its weight will be calculated using (1) and considering $w_i=0$.

After the end of the update procedure, if the length of the keyword list is *len* and *len>user_max* then the *(len-user_max)* keywords with the smallest weights will be deleted. This is done to prevent the expansion of the list and minimize the times that the number of keywords that have been added to the user's profile do not reflect his/her interest.

One critical decision is that the update procedure will take place only for the shows the viewer watched at least t_{MIN} and at the time the show is supposed to end, in order for the system to know the total number of minutes the viewer watched the show.

Figure 6. Update weight function plot (y:(1) x:time)

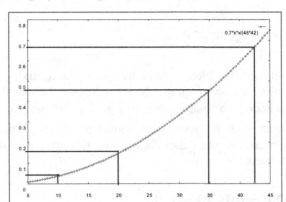

Consider the following example. Assume that viewer X watched a show S with t_d=45 minutes, and the update procedure for the keyword k is taking place. Also, assume that the weight of the specific keyword in the show's S list is $sw_k = 0.7$ and $t_{DAT\_TMA}(z)$= 42 minutes and initially w_i=0. Figure 6 depicts the plot of the weight update function that represents the new values of w_k with respect to the number of minutes the viewer X watched the show S. In the case where t_u=10 minutes the w_i will be increased by less than 0.04, when t_u=20 w_i will be increased by more that 0.15, when t_u=35 w_i will be increased by approximately 0.45, and finally when t_u=42.5 w_i will be increased by more than 0.65.

Updating IMA's Weights

We choose a simple procedure for updating IMA weights. The update procedure takes place each time a user logs out of his/her computer account and takes into account for each Web page the user visited, the weight of its keywords, the time the user spent reading this Web page, and the total time the user spent reading all the Web pages he/she visited during that specific session.[9]

Assume that U is the set of the sites the user visited, each Web page v has a list of the keywords L_v, for each keyword k in L_v has a w^v_k that reflects the importance of the keyword k in the Web page's content.[10] Also assume that the system for each Web page $v \in U$ maintains the time t_v that the window of the Web page was active. Then,

$$\forall\, v \in U$$
$$\qquad \forall\, k \in L_v$$
$$\qquad\qquad w_k \leftarrow w_k + w^v_k * t_v / t_d \qquad\qquad\qquad\qquad (2)$$

where w_k is the weight of the k keyword in the user's profile, and t_d is the total time the user spent reading all Web pages in U. If the keyword is not found inside the user's profile, it is added, considering that its initial weight was equal to 0.

As in the TMA case, after the end of the update procedure, if the length of the keyword list is *len* and *len>max* then the *(len-max)* keywords with the smallest weights will be deleted.

The Case of Multiple Viewers

Consider now the case of multiple viewers watching TV at the same time. The advertisement selection procedure differs from the case of the single viewer only in how the COD agent generates the list that the channel provider requests before a commercial break. Thus, the advertisement selection algorithm that is being executed by the channel provider is exactly the same as described in the *Selecting Advertisement for a Single User* section.

Note that for the case of the multiple viewers, each viewer is associated with a pair of TMA and IMA agents that maintain his/her TV and Internet preferences, respectively.

Figure 7. Merging L_x lists (where x $\in$ {Georgia, John, Nick})

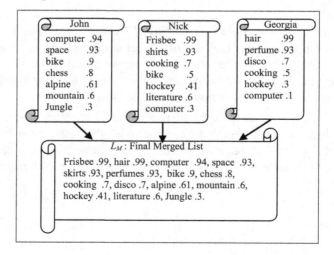

The COD agent operates as follows: for each viewer x that it believes is currently watching TV, it performs the procedure described in 5.2.3 and stores the generated merge list in a separate list L_x. After finishing this procedure, it merges the L_x lists into the final L_m list, which it sends to the channel provider. An example is shown in Figure 7.

Note that we do not have the COD agent directly merging the lists that the TMAs and IMAs send. We want to allow the flexibility for extending our merge procedure in the future to allow some viewer profiles to contribute more than others to the final list.[11]

For the procedure to update weights for each of the viewers the systems believes are watching TV, their keyword list is updated by following exactly the same procedure as in the case of the single viewer.

As we saw in ADVERTISEMENT SELECTION algorithm, the channel provider can determine how much the viewer's preferences and the show's keyword list will contribute to the advertisement selection by setting the k_{COD} and k_{SHOW} parameters. In the case of multiple viewers, a rational tactic would be for the channel provider to consider more heavily the show's keywords than the viewers' profiles.[12]

Not Knowing Who is Actually Viewing

In cases where we know with certainty exactly who is watching a show and for how long they are watching, we can make use of detailed user profile information to customize the ads being presented. In some cases, however, this is not easy to determine.

Several current approaches that monitor viewers' preferences assume that only one viewer each time is watching TV (e.g., Lekakos et al., 2001). This assumption is made mainly because even for one person it is very hard to find his/her ID, thus finding the IDs when more than one person is inside a room is a harder problem to solve. Other systems just assume that it is somehow known with high probability who the person is who is watching TV at any point in time.

To determine who is watching we could use mechanisms such as voice analysis and statistical information on who is more likely to be at home at any time of day for each day and what the viewer is more likely to be watching on TV. For voice analysis, the system can compare the voices in the room with the voices of the people in the house that are stored in its database, while at the same time it can recognize the names and words like "dad," "father," "mother," "mum," and also the names of the persons that live inside the house.

But these approaches have problems. A viewer may be in the room but not talking or talking but not watching TV. There may be too many people in the room for the voice recognition software to work properly—requiring users to register when they view removes the spontaneity of TV watching. And statistical information may not allow for new viewing habits.

We believe that the possibility for "wrong assumptions" can in fact be used to an advantage. We want to make use of the fact that advertisers want not only to direct their ads to viewers who are already interested in their products but also to generate new interest among viewers.

Because of this, we want a viewer's profile to not only cover the areas in which he/she is interested, but also to be broad enough to satisfy advertisers' needs for generating new interest. The idea is to expand viewers' profiles by adding keywords from shows that he/she watched, even if it turns out that they only watched the show for a very short time. Thus, with voice recognition software operating, anyone who is identified as being in the room can, to begin with, be assumed to be watching and to be watching for the duration of the show.[13] This more liberal interpretation will, at the very least, allow advertisers to try to pitch their products to these viewers.

Our model will adjust its viewer profiles in this kind of liberal manner, as follows: consider the case where John and Nick are watching TV, and Georgia just stepped in the room for couple of minutes. The list that the COD agent will send to the channel provider is the one in Figure 7 and is the result of the combination of the L_x[14] lists of viewers Georgia, John, and Nick. In fact, the profiles shown in Figure 7 would be used if all three of the viewers were initially detected in the room and there were no direct reasons for assuming that anyone had left. Thus, Georgia's preferences will affect the ads that Nick and John will watch, in this way broadening the range of potential ads for Nick and John.

To possibly adjust assumptions about viewers continuing to be in a room, there are some heuristics that could be used. One approach for example is to have the system update the keyword list of a viewer for t minutes from the last time it detected the viewer inside the room, where t is equal to the average time the viewer usually watches TV that specific day. In our example, if we assume that the day is Monday and Georgia usually watches TV about 45 minutes on Monday, it will update for 45 minutes her profile according to the shows John and Nick were watching.

Future Work

As outlined in Pramataris, Papakyriakopoulos, Lekakos, and Mylonopoulos (2001), iTV is a new medium, allowing viewers to browse and select shows all at once. Users can be initiating interaction and the interface choices must then be simple. Next, we suggest two ways of introducing more interactivity from users, followed by a brief discussion of how to experiment with alternative calculations in parts of the model.

Allowing Viewing on Demand

One very interesting proposal for future work is to introduce some flexibility for users to watch on-demand shows (or part of the shows) they missed. In order to enhance our system with extra functionalities and to make it more competitive and attractive to channel providers, we suggest the *FEtch Time UNIT*, or more simply the *FETUNIT*. The FETUNIT gives the viewer the ability of watching, on demand, the parts of a show that he/she missed by switching between channels. The idea is:

1.   Provide the viewers with a limited number of free FETUNITs when signing up.
2.   Each time a viewer wants to watch t_{missed} minutes of a missed program he/she loses t_{missed} FETUNITs and the relevant keywords in his/her profile are updated,[15] using (1), where $t_u = t_d$.
3.   User can buy FETUNITs or exchange them with more ads. The idea of the FETUNIT is based on maximization of the channel provider's profit by selling more advanced services to viewers.

Allowing Bookmarking of the Shows

Another advantage of our approach is that it can be easily expanded to support other valuable features. More particularly, Ardissono et al. (2003) proposed a bookmarking option. This could be easily applied in our approach as follows: each time a user likes a show he/she should be able to bookmark it. Then, even if the viewer only watched a part of the show the system could update the user's weight just as in the case where the viewer had watched the whole show. The viewer could also have the ability of bookmarking a show that he/she has not watched, by using the online program schedule screen. The reasons are either to have the system notify him/her the next time the specific show will be transmitted or for using an *on-demand* service to see the missed show at a later time.

Investigating Alternative Formulas

We have already suggested a few places where alternative calculations may be tried within our model. For example, concerning the combination of the TMA and IMA lists, there are a number of different policies that can be used. For instance, instead of considering the weight with the highest value we could consider the average weight value. For future work, we could simulate our model working with different possible formulae to compare the relative value of these different calculations.

We can also explore further how to exploit information about distinct day preferences. For instance, many people have preferences for viewing tied to the day of the week, for example, watching movies on Friday rather than regular TV shows or watching TV from Monday to Friday 6:00 a.m. to 7:00 a.m. (before they go to work) but not watching during this time on the weekends. Knowing more about a viewer's day preferences can help both to infer who is most likely watching a particular show and also to deliver ads customized to classes of viewers with a similar kind of viewing profile.

In addition, we would also like to allow users to have a more active role in adjusting some of their profile information. Allowing users to view how they are represented by the TMA would, for instance, make it possible for them to adjust the weights used within the calculations.

Another very interesting issue to study is to find a function that, considering the number of viewers, would be able to determine the optimal k_{COD} and k_{SHOW} values.

Conclusion

In this work we focused on a new area of personalization in iTV, and more specifically we present a mechanism for personalizing advertisements. Our goal was to provide users with cheap TV services that have less annoying advertisements. To do this, we proposed an approach based on monitoring viewers' behavior while watching TV and surfing on the Internet.

In summary, our approach: (1) offers users cheap TV services with advertisements that the viewer might be more interested in watching, thus making him/her less annoyed, (2) maximizes the advertisements' effect by associating them with specific shows that might contain the advertised products, and (3) is concerned with the problem of recognizing the people that watch TV but suggest this can be turned to an advantage, in order to expand a viewer's profile and to make things more attractive to the advertiser.

Our proposal to begin with a keyword-based representation of all possible ads to be aired with a show is also considered in the work of Thawani et al. (2004). But we have taken this suggestion a step further by outlining a specific advertisement selection algorithm that makes use of information about the current show and the user's past activities.

We also propose to save on network broadcast bandwidth by processing only those ads that are to be aired for a particular show. Our approach, however, could be easily modified to support broadcasting by labeling all the broadcast advertisements with the IDs of the set-top box that should be displayed. In other words, instead of the channel provider sending to each set-top box the advertisements that the users in that house are going to watch, it could attach to each of the advertisements the list of the set-top box that should be displayed as well as their sequence.

Researchers in the field of user modeling have been investigating for many years now how best to employ user profile information to direct the reasoning of intelligent systems. We present an approach that allows viewers' preferences to influence the selection of ads. Without such personalization, the choice of ads would need to be based on a stereotype (as in Ardissono et al., 2003) or would simply need to appeal to a generic user. The promise of personalization, as captured in our model, is to make both users and advertisers more satisfied, finding ads most appropriate to the viewers.

We can learn as well from some of the general lessons about personalization discussed by Weld et al. (2003) who claim that *"effective personalization requires improved methods for both adaptation ... and customization."* In our research, we are indeed concerned both with adapting, by dynamically adjusting our user profiles and ad space, as well as customizing, by considering methods to be sensitive to the profiles of the viewers.

References

Ardissono, L., Gena, C., Torasso, P., Bellifemine, F., Chiarotto, A., Difino, A., et al. (2003). *Personalized recommendation of TV programs.* In Lecture Notes in Computer Science, Volume 2829, AI*IA 2003: Advances in Artificial Intelligence: 8th Congress of the Italian Association for Artificial Intelligence Pisa, Italy, September 23-26, 2003 Proceedings (pp. 474-486). Springer Berlin.

Kurapati, K., Gutta, S., Schaffer, D., Martino, J., & Zimmerman, J. A. (2001). *Multi-agent TV recommender.* Paper presented at User Modeling 2001: Personalization in Future TV Workshop, Sonthofen, Germany.

Lekakos, G., & Giaglis, G. (2002). Delivering personalized advertisements in digital television: A methodology and empirical evaluation. In *Proceedings of the AH2002 Workshop on Personalization in Future TV* (pp. 119-129).

Lekakos, G., Papakiriakopoulos, D., & Chorianopoulos, K. (2001). *An integrated approach to interactive and personalized TV advertising.* Paper presented at User Modeling 2001: Personalization in Future TV Workshop, Sonthofen, Germany.

Masthoff, J. (2004). Group modeling: Selecting a sequence of television items to suit a group of viewers. *User Modeling and User-Adapted Interaction, 14*(1), 37-85.

Merialdo, B., Lee, K. T., Luparello, D., & Roudaire, J. (1999, October 30-November 5). Automatic construction of personalized TV news programs. In *Proceedings of the Seventh ACM international Conference on Multimedia (Part 1)* (pp. 323-331). New York: ACM Press.

Pramataris, K., Papakyriakopoulos, D., Lekakos, G., & Mylonopoulos, N. (2001). Personalized interactive TV advertising: The IMEDIA business model. *Journal of Electronic Markets, 11,* 17-25.

Thawani, A., Gopalan, S., & Sridhar, V. (2004). Context aware personalized ad insertion in an interactive TV environment. In *Proceedings of the AH2004 Workshop on Personalization in Future TV* (pp. 239-245).

Weld, D. S., Anderson, C., Domingos, P., Etzioni, O., Krzysztof, G., Lau, T., et al. (2003). Automatically personalizing user interfaces. In *Proceedings 18th International Joint Conference on Artificial Intelligence, IJCAI03* (pp. 1613-1619).

Xu, J., Zhang, L., Lu, H., & Li, Y. (2002). The development and prospect of personalized TV program recommendation systems. In *Proceedings of the IEEE Fourth International Symposium on Multimedia Software Engineering (MSE02)* (pp. 82-89). Piscataway, N J: IEEE.

Endnotes

[1]   Stereotypical information is also used; we do not focus on this kind of knowledge for our work.

[2]   We focus our discussion in this chapter on how to implicitly acquire the viewer's profile information.

[3]   Keywords with the highest value will reflect what the user really likes. Including keywords with the lowest values will be used to allow advertisers win over new users in a household of multiple users. See *The Case of Multiple Viewers* section for further discussion.

[4]   There are a number of different policies that can be used for the combination of the TMA and IMA lists. We plan to explore different policies for future work.

[5]   The channel provider can specify the percentage of keywords to come from each of the CODs and the show's lists and multiply them by k_{MAX} to generate k_{COD} and k_{SHOW}.

[6] $L_{ads}(p)$ is the list of the keywords and their weights that describe the advertisement $p \in P$, k_{COD} is the number of the keywords that will be used from the L_{COD}

list, k_{SHOW} is the number of the keywords that will be used from the L_{SHOW} list, $t_{ads}(p)$ is the time duration of advertisement p, t_{ad} is time duration of ads, and t_{min} is the minimum time duration of the ads in the system's database.

[7]   Note if the preferred ads do not completely fill the commercial break time, the remaining time can be filled with ads for upcoming shows or public service ads.

[8] $w_i \leftarrow w_i + sw_i * (t_u / t_d) * (t_u / t_{DAT\_TMA}(z))$

[9]   By the term session we refer to the time between the user's log in and the user's log out.

[10]   For now we will consider that w^v_k is always 1. Greater use of this possible weighting factor is left for future work.

[11]   For future work we can study whether it is useful sometimes to weight one viewer's list more heavily. It is also possible to use different criteria for merging the list as discussed in Masthoff (2004).

[12]   Note that the show's keywords reflect at least the current interest of the viewers.

[13]   If it is not possible to know who is in the room, the system can simply attempt to cater to all members of the household. Note that even if we know who is in the room we do not know for certain who is watching TV.

[14]   By the notation of L_x we refer to the merged list that is the result of the combination of the TMA and IMA list according to the procedure described in the *COD Agent* section.

[15]   Even if the viewer's profile has been updated according to the time he/she watched the show, it is rational to update it again considering this time that he/she watched the whole show, because spending FETUNIT is an indicator that the viewer is really interested in the show's content.

Chapter V

Analysis and Development of an MHP Application for Live Event Broadcasting and Video Conferencing

Kristof Demeyere, Ghent University, Belgium

Tom Deryckere, Ghent University, Belgium

Mickiel Ide, Ghent University, Belgium

Luc Martens, Ghent University, Belgium

Abstract

This chapter will introduce a technology framework that can be used to add video conferencing services and live video events on the multimedia home platform (MHP). The solution is based on a bridge between Internet protocol (IP)-networks and digital video broadcasting (DVB) channels in order to stream video that originates from an IP network into the broadcast. The introduction of (iDTV) is completely changing the user experience of television in the living room. In our opinion, the iTV infrastructure lends itself perfectly to support live event broadcasting and video conferencing, both enriched with interactive applications. These services have a vast application domain which includes plain video conferencing but also video surveillance, t-learning, t-health, and user-centric content services. The objective of the framework is to provide basic functionality to the service provider to create these and other innovate services.

Introduction

Interactive digital television (iDTV) is gradually replacing his analogue predecessor, whom everyone has been so familiar with for a long time. iDTV offers superior image and sound quality. On top of that, it enables the user to interact with the broadcast ed services. The magic box that unlocks the wealth of new services is the set-top box (STB). An STB is capable of decoding the broadcasted digital video signal and provides local as well as regional interactivity. Next to the introduction of iDTV, there is the growing popularity of multimedia video services on the Internet. Live events are being broadcasted, people share their own video files, and communicate with each other using low cost webcams. In this chapter we will give an overview of technologies that will enable the convergence of these advanced video services from PC to the MHP. We will start with an overview of relevant and existing technologies together with their limitations. Finally, we will propose an architecture for live event broadcasting and video conferencing.

MHP

The multimedia home platfom (MHP) is the first common open middleware platform being deployed at a large scale in STB. It provides the application developer with a number of application program interfaces (API's) suitable to develop a variety of applications irrespective of the underlying hardware of the STB. Applications developed for this MHP will be interoperable with different MHP implementations resulting in a horizontal market. The MHP standard considers the following profiles: *enhanced broadcast profile, interactive broadcast profile,* and *Internet access profile.* The first two profiles are incorporated in the MHP specification 1.0 while the last one is present in the more recent specification MHP 1.1. Each profile refers to an application area and describes the required STB capabilities. Enhanced broadcast profile combined digital broadcast of audio and video with applications providing local interactivity and does not require the STB to have a return channel. However, the interactive broadcast profile extends the enhanced broadcast profile by enabling services with regional interactivity, which requires the presence of a return channel. The Internet profile, at last, extends interactive broadcast by providing Internet services and requires a more sophisticated STB with more processing power and memory. Most of today's STBs labelled MHP compliant, currently implement MHP specification 1.0.2 or 1.1.2 (European Telecommunications Standards Institute [ETSI], 2002, 2005).

Live Event Broadcasting and Video Conferencing

Live event broadcasting and video conferencing both require regional interactivity, meaning that at least the interactive broadcast profile has to be supported. With both services, custom audiovisual content, as requested by the user, has to be delivered to the STB. This audiovisual content can range from the video from a participant in a video conference to a specific camera angle video stream resulting in a wealth of application scenarios. In general this audiovisual stream will originate from a IP-network like the Internet. The use of the Internet as originator of the content will make publishing video content accessible for regular users and creates some interesting opportunities for new user user generated content television. Looking at a typical STB there are two ways to receive audiovisual content: the return channel and the broadcast channel. The return channel provides a bidirectional communication link between the STB and an IP network, while the broadcast channel is a unidirectional channel that pushes data to the STB. Given the flexibility by which each user should be able to start a video conference or select a live event to watch, the use of the return channel seems preferable. In that case the return channel should provide sufficient bandwidth to ensure an acceptable video quality. Video conferencing also requires that each participant is able to capture his own video and audio stream in order to transmit this media to other participants.

Application Scenarios

The main purpose of this chapter is to propose a framework that enables new video services for the MHP. To illustrate the need for such a framework, some future applications are described here. Obviously, this framework is not an endpoint, but the foundation for further innovation.

Live Event Broadcasting

Local organisations may have interest in broadcasting live events such as local music festivals where different concerts are being recorded. The user at home can start the MHP application and select in a custom graphical user interface (GUI) the concert that he/she wants to view. In this way, live events can be accessed on demand by people sitting in front of their television.

The same principle can be applied to sporting events. Kindergartens can stream video to the parents or grandparents to their home. In a third scenario, people can stream video from security cameras placed in their houses to friends and family so

Figure 1. High level architecture for live event broadcasting and video conferencing

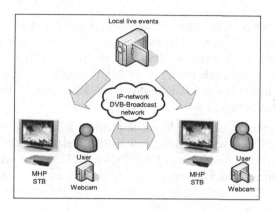

they can survey the house. In this case, valid authentication, with for example an electronic identity card, well be necessary.

Video Conferencing

Two friends can set up a video communication while watching television or the grandparents can communicate with their grandchildren while sitting in their living room. A third scenario imaginable is a t-learning experience where the teacher and student communicate live during the t-learning session.

All these scenarios have in common that they use live video, and that the users expect low cost, common equipment like a television set, STB, webcam, and Internet connection.

MHP Specific Limitations

The MHP specification provides a rich set of features for the application developer to use, but at the same time it lacks support of specific functionality required to implement live event broadcasting and video conferencing. In the following sections we illustrate the lack of functionality to bring on-demand content to MHP STBs in a cost efficient way.

Although the application developer is bound by the features provided by the API, it is worth taking into consideration the restrictions of current STBs.

Video Delivery Through the Return Channel

The return channel is the path through which the STB can communicate with the outside world. Basically, this return channel can take any form, ranging from a low-speed Internet connection through a PSTN modem or GPRS, to a high-speed Internet connection through a cable modem. The MHP specification defines a set of interaction protocols accessible to MHP applications depending on the supported profile. Figure 2 illustrates this protocol stack (ETSI, 2002).

It should be remarked that most STBs do not implement all of these protocols. To stream live video to the STB it seems sufficient to use the real time transport protocol (RTP) that relies on user datagram protocol (UDP) over IP. However, when it comes down to using this return channel for receiving content such as video or audio, there are certain limitations. Delivering streaming media through the return channel is currently not supported by MHP (ETSI, 2002, p. 52). It is however interesting to note that some STB manufacturers have equipped their boxes with this functionality. The MHP standard can be improved by enabling playback of audiovisual content delivered through the return channel, which would take the MHP standard closer to internet protocol television (IPTV). MHP for IPTV has been proposed (Piesing, 2006) and is currently in process of standardisation.

There are two possible ways to work around the restriction. One way is to render successive frames of a video sequence in the background layer and a second way is using the graphics layer of the MHP application for displaying video frames.

When displaying the video in the background layer, the video has to be encoded as Motion Picture Experts Group (MPEG)-2 I- and P-frames. These frames can be retrieved from the return path and periodically displayed on the background layer. For displaying video on the graphics layer, the video has to be encoded as gif, jpeg,

Figure 2. Return protocol stack

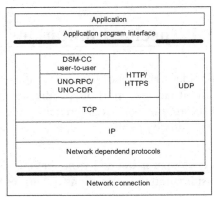

Figure 3. MHP display layers

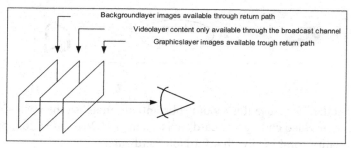

or png images. These images are also available on the return path. Due to bandwidth limitations and latency in the network these are not appropriate ways to deliver video at an acceptable frame rate. Both methods are likely to fail in delivering an acceptable video quality. Moreover, this approach forces the separation of audio and video which might result in synchronisation problems at playback. Extending the java media framework (JMF) 1.0 API (Sun Microsystems, 2005) included in the MHP specifications could be another approach. The JMF API provides functionality to control how media is to be decoded and displayed. DataSources, as described in the JMF API, fetch media data and deliver it to a player which decodes and presents the media. Providing a custom made DataSource could make it possible to receive media through the return channel. Unfortunately, such extensions are not supported.

Since the return channel is not available for streaming video and audio delivery, we will propose an alternative by transferring the streaming media coming from a packet-switched network into the broadcast. This, however, requires that the content is correctly signalled in the broadcast, and the STB is informed where to look for that specific content. This approach is used here for developing the framework.

Peripherals for Capturing Video

In the current MHP specifications there is no other devices for the user to interact with but remote control and keyboard. STBs now available on the market have only a limited amount of support for peripherals like keyboards. STBs with video camera support are rarely available, but it is a required feature for video conferencing. Although there is currently no support for such capturing devices, it is expected that future MHP specifications and STBs will support numerous peripherals among which for example a universal serial bus (USB) webcam. It should be mentioned that if an STB provides a connector for a capture device, the STB will have to deal with the encoding of the captured media in a format understandable for the intended receiver. It is unlikely that STB manufacturers will equip their STBs with advanced

Figure 4. Connecting the webcam and STB in the home network

hardware encoders, because this would substantially increase the cost of the STB unit. Yet, several video coding standards exist such as H.261 and H.263, providing acceptable quality combined with affordable hardware.

The absence of a capturing device can be solved by using intelligent peripherals incorporated in a home network together with the STB. The incorporation into the home network could range from basic IP connectivity to full integration into an existing home network using for example, open services gateway initiative (OSGi) or universal plug and play (UPnP) technology. Some scenarios based on MHP and OSGi describing the interaction between digital TV receivers and home networks have been illustrated (Tchakenko & Kornet, 2004). The open cable application platform (OCAP) already provides an extension for home networking (CableLabs, 2005). Inside the DVB MHP group, the MHP Home Networking (MHP-HN) group has addressed this topic among others.

The solution we have chosen is based on incorporating the STB in the home network simply by connecting it to the residential gateway. In the home network a PC with a webcam is installed to capture the video. The home network and the use of the PC must be transparent to the user who only interacts with the MHP application on the STB. Since no implementation of OSGi or UPnP was available, we created a protocol called the discover configure command protocol (DCCP). This simple protocol is based on the dynamic host configuration protocol (DHCP) (Droms, 1997). The protocol consists of a range of messages used to discover the available webcams on the network and choose one. Control messages are also available to start, stop, and configure the webcam.

Architecture Exploration

The following paragraph briefly discusses several possible architectures able to deliver the video conferencing and live event services. Some of the architectures rely on special STBs or use features that are not included in the MHP specifications. Nevertheless, it is interesting to mention those solutions as they are likely to appear on the market soon.

Embedded Solution

STB manufacturers are incorporating delivery of media over IP in the STB design (so-called IP STB) or extending current STBs with special application specific integrated circuits (ASIC) modules. In this case the architecture for any video delivery service such as video conferencing or live event broadcasting is quite straight forward. The STB can be directly supplied with the media content it requested by streaming to the STB through the return channel. In this case a peer-to-peer (P2P) architecture seems self-evident.

IP Triple Play in HFC

Some STB vendors have announced STBs provided with the functionality to receive two signals through two separate tuners. One tuner will be used to receive the broadcast MPEG-2 Multi Program Transport Stream (MPTS), while the other one will be used to receive the (Euro-)DOCSIS® downstream (Bugajski, 2004, p. 5). These STB, labelled dual feed STBs, will be able to play media content delivered through (Euro-)DOCSIS® by using the same MPEG-2 decoder as used to process the broadcast. Delivering media content to the STB in this case again requires only streaming the content to the STB using its IP address. In the back end the CMTS then inserts the media in the broadcast where the STB can receive this by using a tuner to process the (Euro-)DOCSIS® downstream. A P2P architecture for this approach seems the right choice.

Both cases illustrate the migration to IPTV and enable many new applications including video conferencing, video on demand (VoD), and live event broadcasting. Embedded solutions may have greater flexibility because their range of supported video formats may exceed that of dual feed STBs, which could be forced to use the MPEG-2 decoder to process content.

MPEG-2 Transport Stream Pipe

Although the solutions mentioned previously have a large potential, most of the deployed STBs do not have the ability to receive streaming media over IP. The developed MHP application is intended to run on currently deployed STBs. For that reason the media content requested by the STB will be inserted in the broadcast stream and accessed by the STB in about the same way as it always does to view a service. Assuming that the media is located somewhere on a networked host, we

Figure 5. Bridging video from the IP network to the broadcast network

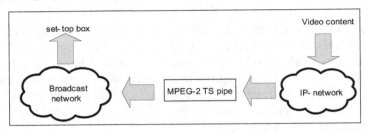

need a so-called MPEG-2 TS stream pipe between the IP and the broadcast network. This component is shown in Figure 5.

Delivering Video on the Broadcast Network

In order to receive video on the MHP STB, the broadcasted data must conform to the DVB specifications regarding video broadcasting. This is based on the MPEG-2 video specification where MPEG-2 encoded video and audio is transmitted in 188 bytes MPEG-2 transport stream (TS) packets. When only one program is available in the TS, the stream is called a single program transport stream (SPTS). When multiple programs are available, we talk about a multi program transport stream (MPTS). Together with the video and audio, information tables are multiplexed in the TS, which give information about the structure and the content of the TS.

Information of the physical network is given in the network information table (NIT), which contains the identifier of the network (Network ID). The program association table (PAT) contains the list of available services in the transport stream and the identifier of the TS (TSID). Each service also has a unique identifier (SID). The three IDs uniquely identify a certain service in the broadcast and are used to create the DVB locator. The DVB locater takes the form of the following URL: DVB://ONID.TSID.SID. It is important that each service has a unique locator in order to avoid collisions in the broadcast.

IP2DVB Gateway

To establish the MPEG-2 TS pipe between an IP network over which the media is delivered and the DVB network, an IP2DVB gateway software module is developed. The software module runs on a host equipped with a DVB-ASI PCI card. Media content is streamed to this host over RTP/UDP and the gateway module subsequently feeds this content to the DVB-ASI card. The ASI output will eventually be multiplexed with other DVB streams by a hardware MPEG-2 TS multiplexer (TS Mux) in order to form a valid MPEG-2 MPTS. The content at the input of the IP2DVB gateway can be delivered in any format of choice providing great flexibility at the expense of increased complexity due to transcoding. However, in the architectures for video conferencing and live event broadcasting we assume that the media content being delivered is already coded as an MPEG-2 audiovisual stream and encapsulated in an MPEG-2 SPTS, relieving the gateway module of the transcoding task.

In order for the STB to be able to find that specific inserted content it needs a DVB locator pointing to the service containing that video content. A DVB locator consists of the original network ID, the transport ID, and the service ID of a service. These IDs inform the MPEG-2 decoder where to look for the service containing the video stream in the broadcast. When multiple MPEG-2 SPTS are combined to form an MPTS, an operation referred to as multiplexing, there can be a remapping of those ID values in order to avoid collisions at the output of the multiplexer. Either it should be guaranteed that there is no remapping of the PIDs in any of the streamed MPEG-2 TS streams containing video, or there should be some form of signalling to inform the receiver about that remapping. In the first case, the originator of the video service can inform the MHP application on the STB about the DVB locator containing the video service since the locator does not change during multiplexing. However, this also means that there has to be some entity in charge of assigning PID values for the various content deliverers to use. This is where the centralised IP2DVB scheduler comes into play. In the second case we need a way to inform the MHP application on the STB of the remapping that takes places during the multiplexing.

The proposed architecture is centralised where the IP2DVB scheduler assigns locators to the content providers and relies on two central entities: the IP2DVB gateway and scheduler. Both will be used in the architectures for live event broadcasting and video conferencing. This architecture has several advantages. First of all, it is compatible with current MHP compliant STBs and as such imposes no additional costs for the customers. Secondly, for broadcasters it requires a minimal investment

Figure 6. MPEG-2 TS pipe

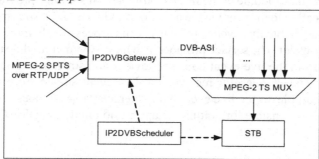

cost since it is an almost complete software solution. Furthermore the system is quite flexible: Services containing requested media can be added and removed dynamically, making it ideal for video conferencing. An important issue for this architecture is resource dimensioning in terms of necessary TSs or services for media-like live video streams. Taking into account the number of users connected to a headend and presuming a certain blocking probability, cable operators can estimate the number of required channels by applying the B-formula or C-formula of Erlang (Hills, 2002; Parkinson, 2002) in the same way a telephone operator does. The latency of the proposed architecture will depend on the performance of the IP2DVB gateway. It should be noted that there are hardware replacements for the IP2DVB gateway available that can reduce the latency at the expense of less flexibility. In the next sections, we reuse the MPEG-2 TS pipe and insert it in an architecture for live event broadcasting and video conferencing.

Architecture for Live Event Broadcasting

There are several requirements for this architectural:

- The user must be able to choose a live event from list of available events.
- Each live event could contain different streams (for example different camera angles).
- The delay has to be as low as possible.
- Content has to originate from an IP network like the Internet.

Figure 6 shows the architecture that can be used for live event broadcasting. It illustrates the functional components as well as the different parties involved in the service delivery. Four main parties can be distinguished. The *live event provider* is responsible for the live event content and publishes details about the event to the *live event aggregator* module of the *service provider*. This service provider delivers the MHP application to the user via the network provider and is also responsible for presenting the available live events to the users and processing their requests to see a particular live event. The service provider also has a streamer module that fetches the live event from the live event content provider and streams it to the IP2DVB gateway. The *network operator* receives the live event stream from the service provider and multiplexes the live event with other available services. The network operator is also responsible for putting the application in the broadcast.

Figure 7. Architecture for live event broadcasting

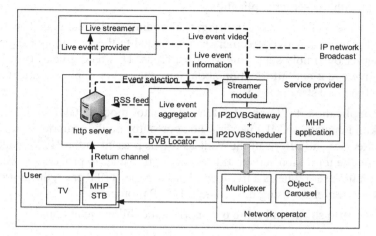

Event Publication: VodCasting

Before the user can choose to view a live event, he or she has to know about the currently available event. To publish such information the usage of VodCasts is suited. Podcasting is a popular way to advertise available audio content create audio content, place it on a server host, and publishing the presence of the content by using PodCasts (which are an extension of really simple syndication [RSS] feeds). A so-called aggregator collects information about available content, informs the user, and fetches the content on request of the user. Using the same principle, one can advertise video content, also known as VodCasting. The RSS feed used is based on the RSS 2.0 specifications with some adaptations. The RSS feed consists of different channels, where each channel represents a separate live event and contains different child elements describing the event. Each channel also consists of one or more items, which contain a VodCast ID that will be used to identify the live event stream. The service provider links each VodCast ID with an available live event.

The server application comprises everything needed to make the live event available to interested users. This application consists of a regular Web server and a communication module. The Web server can be queried by the MHP application. It answers to hyper text transfer protocol (HTTP) queries with the URL of the service provider containing anchors to VodCast feeds describing the live events. In case a live event is requested, the communication module negotiates with the IP2DVB scheduler to obtain the necessary information to start the transcoding and streaming. This communication involves receiving the IP address of the IP2DVB gateway and negotiating PID values for audio, video, and PMT of the service. Consequently, it provides this information to the live streamer, who controls a capturing device and is registered with the server application.

The live streamer performs the capturing, transcoding, and streaming. The system is plug and play as new live streamers can be added to the system and registered with the server by simply supplying VodCast feeds. This feed gives a description about the live event and information about how the server application can locate and contact the live streamer.

The MHP application itself consists of an aggregator module, GUI, and an extensible markup language (XML) parser. The aggregator module queries a server on behalf of the user and offers all the VodCast feeds present in the HTTP page in a catalogue on the television screen. It uses the XML parser to process the VodCast feeds. When the user requests a live event listed on the HTTP page, the aggregator will query the server for using the standard HTTP protocol.

A typical scenario shown in Figure 6 comprises the following actions:

1. The user inputs a uniform resource locator (URL) for the MHP application to retrieve. The aggregator performs the query and processes the HTTP response message resulting in a catalogue appearing on screen.

2. The user selects a live event to view, which results in the aggregator querying the server for this content using HTTP.

3. The server application negotiates through the communication module with the IP2DVB scheduler before starting streaming the content to the IP2DVB gateway.

4. The server application contacts the live streamer and provides the information it received from the IP2DVB scheduler.

5. The streamer application starts streaming live media to the IP2DVB gateway.

6. The server application sends an HTTP response message to the MHP application containing the DVB locator, which causes the STB to tune to that specific service.

Although for this architecture only the interactive broadcast profile is required, the Internet access profile would truly unlock the potential of VodCasting for IDTV. The architecture described previously can not only be used for live event broadcasting but it can also be applied to provide VoD. There is no restriction to what content can be delivered on demand. The only requirement is that it is transcoded if necessary and inserted in a valid MPEG-2 SPTS before being streamed to the IP2DVB gateway. The transcoding can be done by using open-source software like VideoLAN VLC. This media player provides the necessary transcoding functional-

Figure 8. Screen shot of the live event MHP application

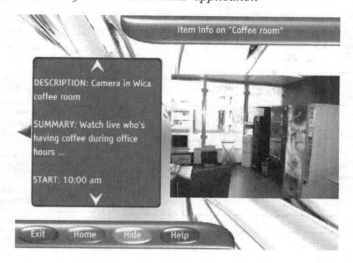

ity in combination with the flexibility to set various parameters like PID values for audio, video, and PMT.

Architecture for Video Conferencing

The architecture for video conferencing is based on Voice over IP (VoIP). The description of the session in terms of parameters like connection or media information relies on the session description protocol (SDP) (Handley & Jacobson, 1998). To initialise a multimedia session between participants, VoIP makes use of the session initiation protocol (SIP) (Rosenberg et al., 2002). By operating on an MHP compliant STB as an SIP endpoint, a video call between a handheld mobile phone and a television set seems plausible. The architecture described next takes advantage of SDP and SIP to describe and initialise a video conference session. The usage of SDP however is limited to passing a DVB locator and other information as explained later on. Figure 8 shows the architecture. Like the architecture of Figure 6, we see here the different functional components and the parties involved. We notice that the content provider is replaced by the user. The IP2DVB gateway and the IP2D-VBscheduler are also in operation.

To handle SIP register requests from users, the architecture has an SIP registrar which is located in the central server. Note that this server also includes a location service database containing the location of the registered users. This server performs the

Figure 9. Video conferencing architecture

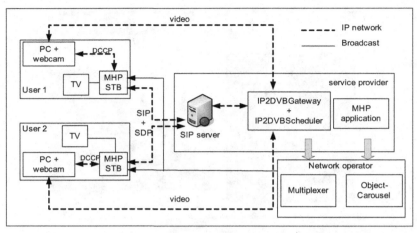

lookup in case of an invitation requested by a user. Before the media session can be initialised, this server also has to negotiate with the IP2DVB scheduler in order to obtain the DVB locator.

The MHP application consists of a SIP module, a network module, and a GUI. The SIP module performs the registration with the registrar, service invitations ordered by the user, and listens for incoming invitations from other users. The network module handles the detection and possible configuration of the capturing devices through the capturing software located on a host in the home network. The capture software module acts as a peripheral connected to the STB. It carries out the capturing, transcoding, and streaming of audio and video. This component could be replaced by for instance an ethernet webcam.

After registration and configuring the capturing module, the scenario depicted in the illustration above consists of the following steps:

1. A user wishes to invite another user for a video conference session. The SIP module issues this command at the central server.

2. The server looks up the online whereabouts of the invited user and forwards the invitation.

3. The SIP module of the invited user informs the user about the invitation, and, if accepted, responds to the invitation with an acknowledgement message.

4. The server forwards this message to the inviting user.

5.   Both MHP applications order the capture software to stream to the IP2DVB gateway and tune to the service containing the live stream sent by the other participant.

The SIP messages are extended with SDP information before being forwarded. This SDP information is limited to a DVB locator and configuration parameters. The locator presents the service containing the live stream from the other participant. The configuration parameters are meant to provide the capturing software with necessary information for streaming. These include IP address of the IP2DVB gateway and the PID values to form the MPEG-2 SPTS. The server extends the acknowledgment message with similar SDP information. Although the server adds SDP information before the addressed user accepts the information, this should not be a problem. On refusal of the invitation, the scheduling can easily be undone when passing the refusal back to the initiating user.

Conclusion

Extending the MHP with new video services like live event broadcasting and video conferencing cannot only improve the user experience of iDTV but has also the potential of introducing new business models that can help the iDTV development. The architecture presented here enables these new services by bringing the flexibility of an IP network and the high bandwidth of a DVB broadcast network together by using a gateway between the two networks. The authors hope that this framework will inspire service operators and content providers to add new services into the broadcast.

References

Bugajski, M. (2004). *Convergence of video in the last mile.* Paper presented at the Jornadas 2004 Conference, Arica, Chile.

CableLabs®. (2005). Open cable application platform (OCAP) specification OCAP home networking extension (Doc. No. OCAP-HNEXT1.0).

Droms, R. (1997, March). *Dynamic host configuration protocol.* Retrieved March 15, 2006, from http://www.ietf.org/rfc/rfc2131.txt

European Telecommunications Standards Institute (ETSI): Digital Video Broadcasting (DVB). (2002, June). *Multimedia home platform (MHP) specification 1.0.2* (Doc. No. TS 101812 v1.2.1).

European Telecommunications Standards Institute (ETSI): Digital Video Broadcasting (DVB). (2005, May). *Multimedia home platform (MHP) specification 1.1* (Doc.No. TS 101812 v1.2.1).

Handley, M., & Jacobson, V. (1998, April). *SDP: Session description protocol.* Retrieved November 20, 2005 from http://www.ietf.org/rfc/rfc2327.txt

Hills, M. (2002, September). *Traffic engineering for the 21st century.* Retrieved March 11, 2006, from http://www.htlt.com/articles

Parkinson, R. (2002). *Traffic engineering techniques in telecommunications.* Retrieved March 11, 2006, from http://www.infotel-systems.com

Piesing, J. (2006, March). *Roadmap for MHP and related technologies.* Paper presented at DVB World 2006—The World of Challenge.

Rosenberg, J., Schulzrinne, H., Camarillo, G., Johnston, A., Peterson, J., Sparks, R., et al. (2002, June). *SIP: Session initiation protocol.* Retrieved November 20, 2005, from http://www.ietf.org/rfc/rfc3261.txt

Sun Microsystems, Inc. (2005). *Java media framework (JMF) 1.0 API specifications.* Retrieved from http://java.sun.com/products/java-media/jmf/1.0/ Tkachenko, D., & Kornet, N. (2004). *Convergence of iDTV and home network platforms.* Russia: St. Petersburg State Polytechnic University.

Chapter VI

Present and Future of Software Graphics Architectures for Interactive Digital Television

Pablo Cesar, CWI: Centrum voor Wiskunde en Informatica, The Netherlands

Keith Baker, Philips Applied Technologies, The Netherlands

Dick Bulterman, CWI: Centrum voor Wiskunde en Informatica,
The Netherlands

Luiz Fernando Gomes Soares, PUC-RIO, Brazil

Samuel Cruz-Lara, LORIA-INRIA Lorraine, France

Annelies Kaptein, Stoneroos, The Netherlands

Abstract

This chapter aims to define a research agenda regarding the software graphics architecture for interactive digital television (iDTV). It is important to note that by iDTV we do not refer to the provision of a return path, but rather to the potential impact the user has over the television (both video stream and applications) content. We can differentiate three major topics to be included in the agenda: (1) to define a suitable declarative environment for television receivers, (2) to research television input (as multiple input devices) and output (multiple display devices) capabilities, and (3) to rethink the models of television distribution and post-distribution (e.g., peer-to-peer [P2P] networks and optical storage technologies). This chapter elaborates on these topics.

Introduction

Digital television receivers are starting to show a reasonable level of maturity and their market penetration is becoming significant (e.g., Italy, Finland, UK, and Korea). A number of multimedia home platform (MHP) (European Telecommunications Standards Institute [ETSI], 2003, 2005) compliant receivers exist and regional standardization initiatives have joined forces by creating the Globally Executable MHP (GEM) standard (ETSI, 2004).

Still, a number of questions regarding the graphics engine (that is, the low-level software presentation control engine) and the interactive capabilities of next-generation receivers arise from both the research community and the industry. Some of the research topics include:

- Definition of a suitable declarative environment for digital television receivers, such as the synchronized multimedia integration language (SMIL) and the World Wide Web Consortium (W3C) recommendation (Bulterman & Rutledge, 2004).
- Integration of other standards, such as Moving Picture Experts Group (MPEG)-4, and Multimedia and Hypermedia information coding Expert Group (MHEG) in current standardization efforts.
- Interaction/visualization using other devices than the remote control/television set (e.g., mobile devices, tablet augmenters).
- Definition of new distribution and post-distribution models, such as P2P and optical storage devices, apart from the typical broadcast model.

This chapter is structured as follows. First, section 2 discusses the state of the art in terms of the broadcast environment, the receiver middleware, and services. Then, based on the state of the art, section 3 identifies relevant research topics in the area. Next, section 4 proposes a research agenda, and section 5 concludes the chapter.

State of the Art

Jensen (2005) has written an interesting study that categorises and defines iDTV services. He differentiates three different iDTV forms:

- **Enhanced:** Enhanced information that is sent via the broadcast channel (e.g., banners)

- **Personalized:** Automatic selection of programs by the receiver (recommendations) and personal digital recording (PDR) capabilities such as play/pause
- **Complete Interactive:** Return channel provision

He points out that, currently, only "low-technology discount solutions," referring to Nielsen usability evaluation methods, are provided. The most important discount solution today is SMS mobile phone return channel, which can evolve in the future to multimedia messaging service (MMS) solutions.

The following subsections describe the state of the art in terms of broadcast environment, software middleware, and services.

Broadcast Environment

Figure 1 depicts a typical example of a terrestrial digital television broadcast system.[1] It is composed of the following components: MPEG2 encoder, digital video broadcasting (DVB) asynchronous serial interface (ASI) internet protocol (IP) link pair, gateway server, remote control/monitor unit, object carousel, multiplexer, modulator/transmitter, and antenna.

Figure 1. Broadcast system (Cesar, 2005)

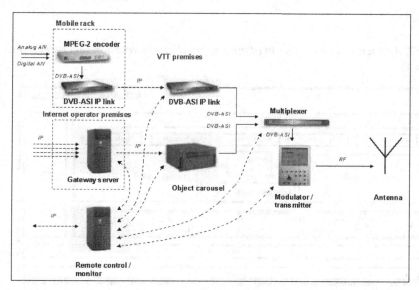

First, the audiovisual stream is encoded with the MPEG-2 encoder. Because people in several locations might encode the audiovisual content, the encoder is stored in a mobile rack and connected to the broadcast system by using the DVB-ASI IP links. The object carousel contains application code and data. It can be uploaded using the gateway server. Next, the multiplexer generates the final MPEG-2 transport stream by combining the audiovisual content (output of the DVB-ASI IP link) and the applications (output of the object carousel). Finally, the modulator/transmitter feeds the multiplexed MPEG-2 transport stream to the antenna. The remote control/monitor can be used to remotely monitor the whole system.

Software Middleware

Third-party development of services requires the digital television receiver to incorporate an interoperable middleware component. Three major regional initiatives exist: MHP[2] in Europe,[3] Advanced Common Application Platform (ACAP[4]) in the USA, and Association of Radio Industries and Business (ARIB[5]) in Japan.

These standardized solutions are composed of a procedural and a declarative environment, as depicted in Figure 2. The two environments do not have to be separated; bridge functionality might link them. In addition to the environments, native applications, proprietary formats, and service-specific software and content can be supported.

The procedural component includes a Java Virtual Machine (JVM) and a set of Java Application Programming Interfaces (APIs), while the declarative corresponds to

Figure 2. Application environment of interactive digital television receivers (Cesar, 2005)

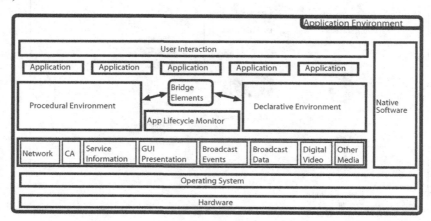

an extensible markup language (XML) user agent. The declarative environment of the three mentioned standards is based on extensible hypertext markup language (XHTML), cascading style sheets (CSS), and television-specific extensions to the Document Object Model (DOM) (Cesar, 2005; International Telecommunication Union [ITU]-T, 2001, 2003, 2004; Morris & Smith-Chaigneau, 2005).

In the case of MHP, the procedural environment is called DVB-Java (DVB-J) and the declarative one is named DVB-HTML. Since the regional standards (MHP, ACAP, and ARIB) share a common approach, DVB has collaborated with the American and Japanese standardization bodies in defining a worldwide application environment: GEM. GEM has been ratified by the ITU in the J.200 (ITU-T, 2001) and J.202 (ITU-T, 2003) recommendations. GEM defines a procedural environment, DVB-J, which includes two Java APIs for graphics: Java Media Framework (JMF) and Home Audio/Video Interoperability (HAVi) (Cesar, 2005; Morris & Smith-Chaigneau, 2005).

GEM defines the television display as three overlapping planes, ordered from bottom to top: background, video, and graphics. The broadcast video is normally hardware decoded and demultiplexed, and then rendered in the video layer by JMF. Alternatively, the applications are developed using HAVi and rendered in the graphics layer, requiring an alpha channel for composition purposes. HAVi extends Personal Java's *java.awt* package, including, for example, remote control events and a television-specific set of widgets.

Figure 3. Worldwide standardization situation of the application environment for iDTV (Cesar, in press)

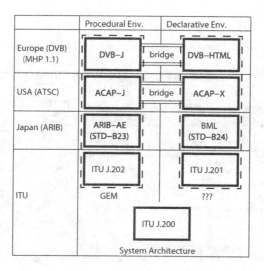

Still, one of the challenges ahead is the definition of a common declarative environment. In order to clarify the goals of the standardization activities, their relationships are depicted in Figure 3.

Apart from conventional digital television receivers such as set-top boxes, companies are producing the next generation of optical storage devices. The two major options are Blu-ray and interactive high definition (iHD). In the case of Blu-ray, its application environment is based on the procedural environment of GEM. In the case of iHD, the application environment is HTML + Time (based on the XHTML + SMIL profile.)

Other technology options for the digital television application environment include MPEG-4, Flash, and MHEG. First, MPEG-4 video codecs are already part of the standards, while the interactive part of MPEG-4 is a popular research topic, but it is not (yet) a real commercial alternative (Cesar, 2005; Baker, 2006).

Flash, on the other hand, is a popular solution for interactive multimedia on the Internet. Still, it has major problems such as: it is a non-structured technology, it uses a scripting language, it is a proprietary format, and dynamic modifications to a presentation are difficult to manage. Finally, MHEG, because of its popularity on digital television in the UK region, is a standard that will be supported by MHP.

Services

The development of MHP applications have become a lucrative business, as can be seen from the number of graphical authoring tools available on the market (e.g., Cardinal Studio[6]). The range of available services is ample;[7] some examples include

Figure 4. Screen shot of Stoneroos program guide (Dutch layout)

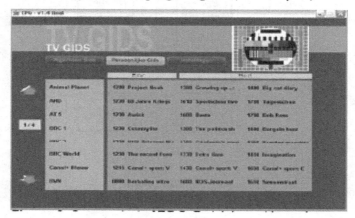

Figure 5. Screen shot of Stoneroos program guide (BBC layout)

games, commercials, home shopping, and banking. Nevertheless, the two major services are the electronic program guide (EPG) and the Super Teletext. The former shows information related to the scheduled audiovisual content. The latter is a new version of the traditional teletext service, in which multimedia content is broadcast within the MPEG-2 stream.

This subsection illustrates current services by taking a look at the work of a Dutch digital television company called Stoneroos. First, we will describe an adaptable EPG and then an interactive application called C-Majeur.

Stoneroos has developed a personalized program guide that uses XML TV. This way, the graphical layout of the application can be adapted. Figure 4 shows the Dutch layout, while Figure 5 shows the British Broadcasting Corporation (BBC) layout.

C-Majeur is a weekly television program about classical music. Stoneroos created an interactive episode of C-Majeur, shown in Figure 6. Viewers could select extra

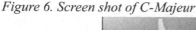

Figure 6. Screen shot of C-Majeur

commentary and a karaoke version of the content in which a scrolling score of the vocals was shown along with close-ups of the soloist and conductor. The interactive extras were available during the entire week following the initial broadcast.

Summary

The following list summarizes the state of the art:

- **Standards:**
    - ◦ **Procedural:** Java-based
    - ◦ **Declarative:** Based on XHTML(+SMIL), CSS, and DOM extensions
- **Services:** A number of selection-based services (e.g., EPG and Super Teletext) plus discount interactive solutions such as SMS voting
- **Current configuration:** Low-print hardware configuration (around 250MHz, 16MB-64MB of RAM, and 8MB-32MB of flash memory), remote control (input), and television set (output). Some models include storage capacity and PDR capabilities (e.g., Tivo, iHD, and Blu-ray devices).
- **Distribution mechanisms:** Broadcast

As a final summary of the current state of the art, Figure 7 depicts the current model of interactive television. The interactive content is broadcasted to the receiver, which includes an application environment. Then, this content is visualized through the television set and the user can interact with the receiver using a remote control. The thickness of an arrow represents the amount of content that is transmitted. The receiver might have a return path connected to a server. Finally, as the figure shows, even though many people can watch the television set at once, only one person can interact with it.

Figure 7. Current model of iDTV

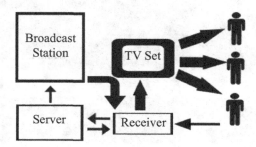

Future of Interactive Digital Television

We can distinguish four different topics on which research should be focused: (1) definition of the concept interactive digital television (IDTV), (2) extensions of current standards, (3) service development, and (4) configuration of digital television receivers. The following subsections elaborate on those issues.

Definition of iDTV

One of the most important questions to be answered by this chapter is: *what do we mean by iDTV? Is it IDTV a dodo of history like the paperless office and the information superhighway? Have we already made of IDTV a meaningless concept?*

In our opinion, IDTV is not about plugging your television to the phone line, to use your mobile phone for voting, or for that matter, to use your normal telephone to call to a television program and achieve your 15 seconds of fame. IDTV is about the potential impact a user has on the digital content. Some forms of interaction include applications that enhance the broadcast content and possibility of selecting different subtitles streams. Higher level of interaction is provided by PDR equipment in the form of pause, play, and skip content (Agamanolis & Bove, 2003) since the user actually affects the status of the content.

Further interactive capabilities include, for example, the possibility to control the content that the user consumes (i.e., personalized television) and end-user enrichment of the incoming material (Cesar, Bulterman, & Jansen, 2006). In our opinion, the research community should focus on the definition of new models that define the iDTV paradigm such as Chorianopoulos (2004) has done with his virtual channel proposal.

Standards

All in all, the definition of the GEM standard is good news for the digital television community. It is a worldwide solution for a digital television application environment. Even though its market penetration is low, apart from the Italian market in which proper subside policies have been applied, its procedural environment is a technology success (e.g., Blu-ray Java intends to reuse it).

Nevertheless, the definition of a suitable declarative environment is a challenge. Currently, apart from the structure and the visual style of the documents, the regional standards define three types of DOM events: user, internal, and broadcast generated (e.g., starting of a show). The major limitation of these standards is that they require the use of scripting for handling these events. For example, in order to synchronize the audio-visual content and an application, a script is activated when a broadcast event is received. This limitation can be overcome by adopting multimedia declarative approaches such as SMIL (Bulterman & Rutledge, 2004) or nested context language (NCL) (Soares & Rodrigues, 2005).

SMIL is the W3C standard for multimedia presentations. SMIL is a declarative format: rather than encoding the functionality of an (interactive) piece of content as a script or within the definition of a particular media codec, SMIL describes the set of interactions among media objects. Then, SMIL allows a particular renderer (or player) to implement this functionality in a manner that best suits the needs of the application runtime environment. The structure of an SMIL presentation follows the simple standard XML form:

```
<smil... >
  <head>
    general declarations for metadata,
    layout and interactive control
  </head>
  <body>
    temporal schedule of contents and
    content events
  </body>
</smil>
```

Any object within a time container (i.e., *par*, *seq*, and *excl*) can have a scheduled or interactive begin time. The events associated with interactive begin times can be triggered by user interactions, by external script objects (via a set of SMIL DOM events), or by events that are triggered within associated content streams.

Hence, the main advantage of using SMIL is two-fold:

1.   SMIL is a well-developed and widely deployed language, available on desktop and mobile devices.

2.   Being declarative, SMIL code is small (compared with Java), and it is easily verifiable—meaning that the risk of introducing viruses and other unwanted interaction side effects is limited.

As shown in Figure 8, SMIL defines 10 major functional grouping elements and attributes (modules). This results in the definition of a number of profiles. Each profile provides a fixed collection of elements and attributes, drawn from one or

Figure 8. SMIL module architecture

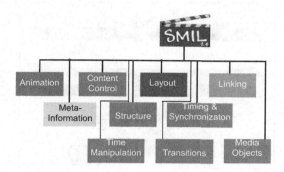

more modules. The purpose of the profile model is to enable the customizing of the integration of SMIL's functionality into a variety of XML-based languages. The major profiles are:

- **SMIL 2.1 Language:** The host-language version of SMIL 2.0
- **SMIL 2.1 Extended Mobile:** A rich language subset for advanced mobile devices
- **SMIL 2.1 Mobile:** A baseline set of features for general mobile devices
- **SMIL 2.1 Basic:** A baseline set of features for low-powered devices with a minimal set of elements and attributes.

In order to facilitate the use of SMIL in an IDTV application environment, W3C is working on two new profiles for SMIL:

- **SMIL iDTV:** A profile containing the major modules to support a broad range of interaction facilities in a lightweight set-top box
- **SMIL Enhanced iDTV:** A profile containing additional enhancements to allow a richer processing across a set of devices

Some of the services these profiles are intended for include stand-alone applications (e.g., photo slides visualized in the television set), semi-synchronized applications (e.g., Teletext with extra audio files and animations), and services synchronized with the broadcast stream (e.g., statistics of a football match).

Figure 9. MAESTRO alone

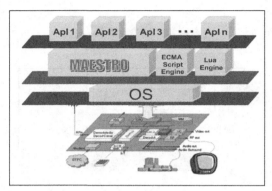

Another declarative solution for IDTV, similar to SMIL, is NCL. NCL claims to do for nonlinear television programs what HTML did for the Web. But, NCL is focused on temporal synchronization of multimedia objects in general and not on user interaction in particular. NCL supports the typical media objects (e.g., video and image), as well as HTML objects.

The Brazilian[8] government is currently standardizing the application environment for their digital television system. One of the alternatives is MAESTRO. MAESTRO is a complete solution, based on NCL.

There are several alternatives for the procedural and declarative middleware by using MAESTRO. The first one, shown in Figure 9, is a resident middleware with only the declarative module. Procedural codes could be launched from declarative applications because MAESTRO allows procedural objects in NCL documents.

Figure 10. MAESTRO + procedural middleware

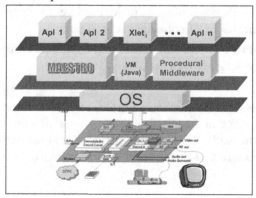

Figure 11. MAESTRO as XLET

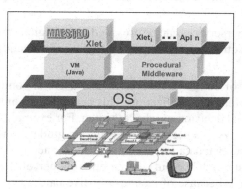

The second alternative, shown in Figure 10, is MAESTRO coupled with a procedural middleware. Of course this alternative will require a set-top box with better performance. However, a better support for all kinds of applications is achieved.

It must be mentioned that both alternatives already presented can run not only NCL but also HTML applications. This means that any content developed for the three main digital TV systems (ARIB, ACAP, and MHP) is supported by MAESTRO.

The third alternative runs MAESTRO triggered by a procedural middleware in Java, as shown in Figure 11. MAESTRO was also implemented as an XLET that can be exported to the user set-top box. Therefore, any application developed in NCL will be able to run in any of the main digital TV systems.

Both SMIL and NCL are interesting proposals for a declarative solution that fits the television model; both are focused on the temporal synchronization of multimedia objects, while HTML was mostly intended for text content. They are different, though, in a number of issues. For example, while SMIL defines strict temporal and spatial semantics (e.g., par and seq), NCL uses templates. These templates can include any kind of semantics.

Finally, another interesting extension to current standardization efforts is to include multi lingual information framework (MLIF) (ISO AWI24616)[9] (Cruz-Lara, Gupta, & Romary, 2005) support. MLIF defines an abstract model to encode the textual information of multimedia services (e.g., subtitles, captions, and t-learning resources). Currently, different groups are working on this issue (e.g., Timed Text in W3C). Hence, MILF does not create a new format from scratch, but deals with the issue of overlap between the formats. The final intention is to create a core framework, in which all the other formats will be integrated. Their research is especially relevant in the case of services serving multilingual communities (e.g., Spain) and in the case of reusing services for different countries (e.g., European services).

Services

A number of innovative services can be provided to the digital television viewers. In this chapter we concentrate on one important topic: recommender systems (Smyth & Cotter, 2000).

Because the amount of content available to end users is continually increasing, there is a need of content personalization. This personalization is a filter and takes place prior to the actual decision of what to watch. For example, Stoneroos has developed I-Fancy. I-Fanzy is a personalized program guide shown in Figures 4 and 5. It uses TV Anytime metadata and its content can be accommodated to different layouts (e.g., Dutch and BBC layout). In our opinion, personalized television is an essential research topic because of its non-intrusive nature and the big amount of available digital content.

Other Distribution Methods

The long unforeseen threat to iDTV is the P2P network. P2P networks have proven to be the killer application of broadband access. These networks are proving to be economically very efficient, and with the BBC leading adoption, this will provide a serious alternative to internet protocol television (IPTV) networks based on client-server technologies.

P2P networks are starting to offer a complete triple play to users for TV, voice, and related IP services. Unfortunately, the type of person-to-person interactivity that a P2P network can sustain is poorly supported by an iDTV platform. If P2P networks based on "Super-peers" are the future of TV, as the BBC's iMP experiment would suggest, the research community has a clear set of challenges to address. These are challenges that can only be addressed from a user-centric viewpoint, commercial and business interests will not design these networks, they will evolve as memes. They will evolve directly in response to user needs, and if not driven by common open innovation platforms, then by the open source communities alone.

One interesting project researching these issues is the I-Share project (Pouwelse, Garbacki, Epema, & Sips, 2005).

Other Devices

In addition to traditional input/output devices, research should be focused on extended user interaction in the television domain. For example, Figure 12 shows two devices already at home that can be used for enriching the television experience.

Figure 12. Interactive devices at home

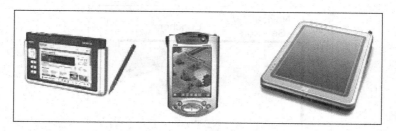

In relation to their output capabilities, one interesting research topic is that of personal devices. Personal devices, apart of incorporating rendering capabilities, are aware of the identity of the user. Thus, these devices (e.g., a handheld device) can personalize television content in a non-intrusive manner. We propose a number of scenarios:

1. **Extra information:** Personal devices can provide enhanced information a user might be interested in, while watching a television program (Finke & Balfanz, 2002, 2004), for example, the statistics of a player during a football match. The key issue is that the information is displayed in the personal device, as thus, it does not disturb the other television viewers in the room.

2. **Audio:** Because the personal device knows the identity of the viewer, it can provide a subset of the content. Imagine, for example, an elderly man who is watching television. He might have lost some of his auditory capabilities. In this case, the personal device might render the audio of the program to the headphones of the elderly person. But, the actual shared volume is maintained at the same level for the other members of the family.

3. **Text:** Similar to the previous example, the personal device might render personally the subtitles of a movie. For example, if I am in a foreign country watching a movie with some foreign friends. I would like to receive the subtitles in my mother tongue, but, not necessarily the rest of the people watching television would like to see those subtitles.

In order to implement the scenarios presented previously, we need a non-monolithic[10] renderer. A nonmonolithic renderer is one that it is capable of deliver parts of a multimedia presentation and extra information to several renderers. The audiovisual content is delivered to the television set. While, user interfaces, extra information, and specific parts of the content are delivered in a synchronized manner to other personal devices.

Figure 13. Classification of devices at home

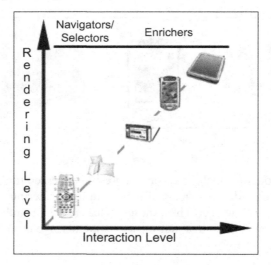

In addition to output devices, other devices can extend the limited remote control interaction provided by current television systems. For example, gesture recognition and voice interaction (Berglund, 2004; Berglund & Johansson, 2004) are interesting topics that need to be further researched.

Figure 13 shows a categorization of devices depending on their rendering and interaction capabilities. We can differentiate two levels of user interaction, when watching content:

- **Navigation/selection.** This level of interaction allows the user to navigate and select content (e.g., go to the next scene).

- **Enrichment.** This level of interaction allows the user to produce his/her personalized copy of the television content. Some examples of this augmentation includes subsetting the content (e.g., deleting scenes) and adding new content (e.g., to include your own audio commentary of a movie).

One example of navigation/selection is ambient technology. We can use everyday objects at home, such as an intelligent pillow, to gather information about the user. These devices can gather context-based information such as the level of excitement of a user. Then, the television renderer can act accordingly to the gathered data (e.g., by pausing the television program in case the level of excitement is too high).

Personal devices can be used to enrich content (Cesar et al., 2006). Figure 14 shows a screen shot of content augmentation (in the form of an image) over television

Figure 14. Screen shot of end-user content augmentation

content. In this case, the end user has highlighted one part of the screen (in this case a tie). For example, he could include, as well, some text such as "I would like this tie for my birthday!" Research in this topic is related to social TV and collaborative work. The major requirement for a content enrichment system is that the base content must not be altered (personal copies should be generated as overlapping layers of content). In addition, because television watching is an entertainment activity, the enrichment process should be fast and simple. We have termed this process as *Authoring from the Sofa.*

Apart from the inclusion of media objects, content enrichment includes virtual edits (e.g., alteration of the presentation timeline and exclusion of media objects) and creation of conditional navigation paths. These capabilities are researched within the European Passepartout[11] project. The scenario in this project is called Maxima. Maxima is the content manager in the house. She selects content and edits it for her children.

Agenda

As a summary, the agenda we propose in this chapter for the future research in iDTV is depicted in Figure 15. From the distribution/post-distribution of content, digital television content (and services) can be transmitted using any kind of network (e.g., broadcast, P2P networks, and mobile networks). Then, the user can utilize any kind of device to interact with and visualize the television content. Each of the devices provides different levels of interaction depending on their limitations. For example,

Figure 15. Proposed agenda for iDTV

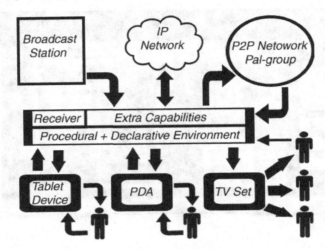

the user can interact with one remote control and one television set (rather limited scenario). But, the user might interact, as well, with a personal tablet augmenter or a personal digital assistant (PDA), non-intrusive interaction is the concept to describe this kind of interaction. Finally, the receiver (and the interactive devices) includes a suitable declarative environment (e.g., SMIL, MAESTRO) and storage capabilities. The receiver might well include extra capabilities such as a recommender engine based on TV Anytime standard and a multilingual framework.

Conclusion

The traditional definition of iDTV as the provision of a return path focuses on the receiver capabilities instead of on the user potentiality. In order to avoid that iDTV becomes a dodo of history, we should move our focus from the receiver to the user. iDTV should not be restricted to one television set and a remote control (intrusive interaction), but extended to other digital devices at home (non-intrusive interaction).

One essential research area is to rethink the validity of the current standards and their evolution. Based on their market penetration of MHP, for example, we might conclude that they are similar to the European Constitution; the parliament of each country approves it, but the citizens are not so sure about it. In our opinion, the problem is in the acceptance timetable of new technologies. Nevertheless, standards should continue evolving (e.g., declarative environment) in order to provide the users the best possible alternatives.

Finally, we should try to model the current situation, in which the traditional role of users as consumers is shifting to become producers (e.g., blogging and personal Web pages) and distributors (e.g., iPod). Much research is needed in topics such as television content enrichment, post-distribution mechanisms (e.g., P2P networks), adaptability of interactive services to other devices, and providing an ambient experience (Baker 2006b).

Acknowledgment

This work was supported by the ITEA project Passepartout and by the NWO project BRICKS.

References

Agamanolis, S. P., & Bove, V. M., Jr. (2003). Viper: A framework for responsive television. *IEE Multimedia, 10*(3), 88-98.

Aroyo, L., Nack, F., Schiphorst, T., Schut, H., KauwATjoe, M. (2007). Personalized ambient media experience: Move me case study. In *Proceedings of the 12th International Conference on intelligent user interfaces* (pp. 298-301)

Baker, K. (2006a). Jules Verne project: Realization of reactive media using MPEG4 on MHP. In *Proceedings of the IEEE International Conference on Consumer Electronics*.

Baker, K. (2006b). Intrusive interactivity is not an ambient experience. *IEEE Multimedia, 13*(2), 4-7.

Berglund, A. (2004). *Augmenting the remote control: Studies in complex information navigation for digital TV*. Unpublished doctoral dissertation, Linkoping University, Sweden.

Berglund, A., & Johansson, P. (2004). Using speech and dialogue for interactive TV navigation. *Universal Access in the Information Society, 3*(3-4), 224-238.

Bulterman, D. C. A., & Rutledge, L. (2004). *SMIL 2.0: Interactive multimedia for Web and mobile devices*. Heidelberg, Germany: Springer-Verlag.

Cesar, P. (2005). *A graphics software architecture for high-end interactive TV terminals*. Doctoral dissertation, Helsinki University of Technology, Finland.

Cesar, P., Bulterman, D. C. A., & Jansen, A. J. (2006a). An architecture for end-user TV content enrichment. In *Proceedings of the 4th European Interactive TV Conference* (pp. 39-47).

Cesar, P., Vuorimaa, P., & Vierinen, J. (2006b). A graphics architecture for high-end interactive television terminals. *ACM Transactions on Multimedia Computing, Communications, and Applications (TOMCCAP)*, *2*(4), 343-357.

Chorianopoulos, K. (2004). *Virtual television channels: Conceptual model, user interface design and affective usability evaluation*. Unpublished doctoral dissertation, Athens University of Economic and Business, Greece.

Cruz-Lara, S., Gupta, J. D., Fernández García, & Romary, L. (2005). Multilingual information framework for handling textual data in digital media. In *Proceedings of the Third International Conference on Active Media Technology* (pp. 82-84).

European Telecommunications Standards Institute. (2003). *Digital video broadcasting (DVB)—Multimedia home platform (MHP) specification* 1.0.3. (Doc. No. ETSI TS 101 812 v1.3.1).

European Telecommunications Standards Institute. (2004). *Digital video broadcasting (DVB)—Globally executable MHP (GEM)*. (Doc. No. ETSI TS 102 819 v.1.2.1).

European Telecommunications Standards Institute. (2005). *Digital video broadcasting (DVB)—Multimedia home platform (MHP) specification 1.1*. (Doc. No. TS 101 812 v1.2.1).

Finke, M., & Balfanz, D. (2002). Interaction with content-augmented video via off-screen hyperlinks for direct information retrieval. In *Proceedings of the International Conference in Central Europe on Computer Graphics, Visualization and Computer Vision* (pp. 187-194).

Finke, M., & Balfanz, D. (2004). A reference architecture supporting hypervideo content for IDTV and the Internet domain. *Computers & Graphics, 28*(2), 179-191.

ITU-T. (2001). *Worldwide common core—Application environment for digital interactive television services* (Doc. No. ITU-T J.200).

ITU-T. (2003). *Harmonization of procedural content formats for interactive TV applications* (Doc. No. ITU-T J.202).

ITU-T. (2004). *Harmonization of declarative content format for interactive TV applications* (Doc. No. ITU-T J.201).

Jensen, J. F. (2005). Interactive television: New genres, new format, new content. In *Proceedings of the Second Australasian Conference on Interactive Entertainment* (pp. 89-96).

Morris, S., & Smith-Chaigneau, A. (2005). *Interactive TV standards: A guide to MHP, OCAP and JavaTV*. Burlington, MA: Focal Press.

Pouwelse, J. A., Garbacki, P., Epema, D. H. J., & Sips, H. J. (2005). The Bittorrent P2P file-sharing system: Measurements and analysis. In *Proceedings of the 4th International Workshop on Peer-to-Peer Systems* (pp. 205-216).

Smyth, B., & Cotter, P. (2000). A personalized television listings service. *Communications of the ACM, 43*(8), 107-111.

Soares, L. F. G., & Rodrigues, R. F. (2005). Nested context model 3.0. Part 5—NCL (Nested Context Language). (Tech. Rep. MCC-26/05, PUC-Rio). Rio de Janeiro, Brazil.

Endnotes

[1] The architecture presented here is the one of OtaDigi. The author collaborated with this project; the intention was to set up a digital television broadcast system for the Otaniemi region, Finland. More information can be obtained in: http://www.otadigi.tv

[2] http://www.mhp.org/

[3] This solution has not been fully supported by the UK and France.

[4] http://www.acap.tv/

[5] http://www.arib.or.jp/english/

[6] http://www.cardinal.fi

[7] Please refer to the MHP Knowledge Project for a complete study on the issue (http://www.mhp-knowledgebase.org/)

[8] It is important to remark that Brazil is the fifth most populated country in the world.

[9] http://www.iso.org/iso/en/CatalogueDetailPage.CatalogueDetail?CSNUMBER=37330&scopelist=PROGRAMME

[10] Our first results in this topic can be found in http://www.ambulantplayer.org

[11] http://www.citi.tudor.lu/QuickPlace/Passepartout/Main.nsf

Chapter VII

Ambient Media and Home Entertainment

Artur Lugmayr, Tampere University of Technology, Finland

Alexandra Pohl, Berlin-Brandenburg (rbb) Innovationsprojekte, Germany

Max Müehhäueser, Technische Universitat Darmstädt, Germany

Jan Kallenbach, Helsinki University of Technology, Finland

Konstantinos Chorianopoulos, Bauhaus University of Weimar, Germany

Abstract

Media are "[media] means effecting or conveying something such as (1) a surrounding or enveloping substance; or (2) a condition or environment in which something may function or flourish; or (3) mode of artistic expression or communication." (Merriam-Webster, n.d.) In the case of ambient media, the humans' natural environment becomes to the "enveloping media" as an environment in which content functions. This work therefore deals with the development of ambient media, far beyond seeing TV as a major entertainment platform in consumers' homes. To satisfy the entertainment-hungry consumer, more and more advanced home entertainment (HE) systems and facilities are required to provide interactive and smart leisure content. This chapter glimpses the future of modern ambient HE systems. Experts in the field of ambient media discuss and contribute to four major lines of the future development of ambient media: (1) social implications, (2) converging media, (3) consumer content, and (4) smart devices.

Introduction

Rather than the consumer explicitly telling a computer system what to do, the system will act autonomously in the way the consumer desires. Natural interaction, personalization, smart metadata, wireless technology, ubiquitous systems, pervasive computation, and embedded systems are technologically enabled. The vision for *ambient media* or *ambient intelligence* has been developed by the European Commission as goal for research projects until the year 2010. The IST Advisory Group (ISTAG) developed the components (see Figure 1) contributing to this vision in their working documents in the beginning of this century (ISTAG, 2001, 2003). The goal of this book chapter is to look far beyond the scope of the utilization of compression techniques for transmitting content or digital television in consumers' homes (Lugmayr, Niiranen, & Kalli, 2004). Ambient intelligence or ambient media seek to make smart technology available for the consumer throughout their natural environment.

Currently many European projects are contributing to this vision with the realization of ambient-media-related projects ranging from digital TV, smart living spaces, media content management, and so forth.

Figure 1. Components of ambient intelligence or ambient media according ISTAG

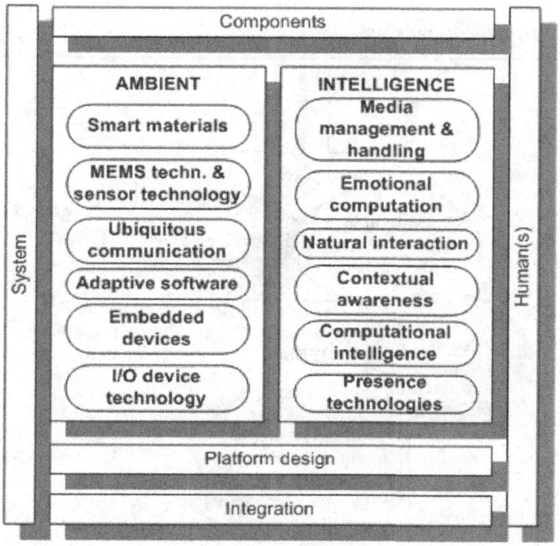

Going further back in the development of multimedia systems, we see the following major development steps in media evolution (defined in Lugmayr, 2003; Lugmayr, Saarinen, & Tournut, 2006):

1. **Natural media:** Forms not requiring electronic technology (e.g., dances, songs, or cave paintings)

2. **Multimedia:** Integrated presentation in one form (e.g., TV, Web pages, interactive installations)

3. **Virtual reality:** Embedding the user into a computer generated world (e.g., CAVEs, computer games, immersive environments)

4. **Ambient multimedia:** The user is exposed to the actual media in their natural environment rather than to computer interfaces (e.g., smart home and spaces, intelligent mobile phones)

5. **Bio-media:** A fully real/synthetic world undistinguishable pure media integrating human capacity (e.g., as aspired by Hollywood films such as Matrix)

Figure 2 gives an overview of the overall chapter topics organized in the following sections:

•   The first section focuses on new technology trends of the ubiquitous computing era fostering substantial changes in all application domains in *Home Entertainment: At the Intersection of Smart Homes & Ambient Entertainment.*

Figure 2. Chapter topic overview

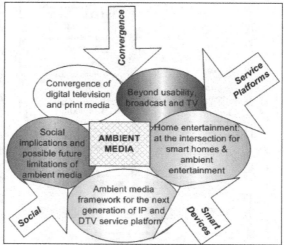

- The second section focuses on media convergence, especially on the future of digital TV with a rather consumer-centered viewpoint in *Beyond Usability, Broadcast, and TV*.

- The third section focuses on the development of design principles for ambient intelligent systems in *Design Principles for Ambient Interactivity*.

- The fourth section deals with the convergence of print media and interactive digital TV (IDTV) as a possible new channel for print media content in *Digital Television, Print Media and Their Convergence*.

This chapter starts with an introduction of HE and potential future trends in the age of ambient media. The first section is followed by sections showing the application of the theories of ambient media in the context of IDTV. IDTV equipment will be one of the major multimedia HE platforms in consumers' homes. The advance of existing services and convergence is of major concern to enable a future-oriented and consumer-friendly platform for accessing ambient media services. Adding interactivity and the development of a set of guidelines and rules on which level they can be implemented on the IDTV platform is described within the scope of this chapter. It is essential to show how media are mediated from their traditional analogue form to their new digital counterparts. This is especially shown in the last section, where print media and their convergence with IDTV are discussed.

Home Entainment: At the Intersection of Smart Homes and Ambient Entertainment

The post PC era of ubiquitous computing (aka pervasive computing, aka ambient intelligence) provides technology trends that will heavily influence the future of IT application domains. The single most important trend is the *pulverization* of hardware and software, which in turn enables an opposite, that is, amalgamating trends towards large application hyper domains, where the pulverized devices and functions are dynamically federated for various purposes. In the remainder, we will summarize the main threat of technology development, together with examples of the immediate consequences for HE hardware and software. For an investigation of midterm consequences, we will investigate the application hyper domains *smart homes* and *ambient entertainment*. The main statements of the present article are depicted in Figure 2. (For further reading see also Braun & Mühlhäuser, 2005; Pering & al., 2005; Straub & Heinemann, 2004.)

Ubiquitous Computing Technology and Consequences for HE

The convergence of consumer electronics (CE) and IT can be considered almost completed, marking the passing on from generation I (CE devices) to generation II as depicted; thereby, the difference between media centers (a CE term) and multimedia PCs (an IT term) became blurred, HE demands pushed powerful graphics into CE and PCs, and IT technology made PCs and networks available in the living room. The latter trend brought along additional features and tore down protectionist walls (sometimes illegally, cf. content piracy).

Generation III approached with the advent of ubiquitous computing, thereby, more substantial and far-reaching changes occurred. Recent product news show early but clear signs; three devices shall therefore be listed here, all meeting at the concept of "federation-enabled business/entertainment nodes" (FedNodes, for short):

- The Sony Playstation Portable PSP™ was acclaimed for its wLAN and Internet capabilities, making it appear like a versatile micronotebook with excellent graphics and computing power.
- Apple's iPod™ is discovered for serious information dissemination (CRM news, lecture videos, etc.).

Figure 3. At the intersection of smart homes and ambient entertainment

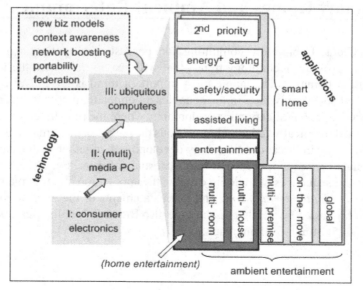

- The Personal Server™ from Intel, a prototype of a wearable Web server with full media capabilities, enters the business-entertainment bridge from the "serious" end.

The latter two, in particular, provide some insight into the role of such nodes in a federated world. The iPod has brought forth an entire industry of peripherals for docking or networking; the Personal Server has, on one hand, been shown with a wrist watch as display and as a twin with a Motorola™ cell phone; on the other hand, it is more or less compatible with Intel's Stargate™ platform intended, among others, as a hub in MICA2™ based sensor networks from Crossbow. In addition, dynamic, zero-configuration federation with ambient peripherals (publicly available wall LCDs, kiosks, car sound systems, etc.) is currently investigated in research (cf. Braun & Mühlhäuser, 2005) and will complement FedNode functionality.

The previous examples substantiate the claim that ubiquitous computing nodes will, in the future, draw much power from being federated, whether or not they serve special purposes; instead of dedicated HE hardware, we will see "HE-inclined" devices, which will be used for other purposes, too, and dynamically federated with various other devices. Major features and consequences are depicted in Figure 3 (upper left): (1) new business models emerge: for example, devices and services can be offered for users on the move, less profitable areas can benefit from hardware, and features "cross-subsidized" from other application domains; (2) context awareness hardware and software can be offered across domains; (3) whenever network connectivity is available/affordable, added functionality can be provided (network boosting); (4) portability and business models have mutual influence and trigger application hyper domains as described later on; and (5) dynamic federation of (rather special-purpose) devices—still a research domain as mentioned—will speed up the trends mentioned before.

Ambient Entertainment

A study by Intel (Pering & al., 2005) shows that wearable entertainment devices lead to a seamless integration of home and nomadic use cases. Video sharing in the digital home (hardly interactive), photograph sharing (interactive in small groups), music sharing (interactive in larger groups), and sharing with the environment spread from very personal to increasingly public areas and introduce new roles, like contributor and proprietor (e.g., a café owner), which complement the provider/consumer stereotype. In a more general perspective, we see that HE scenarios have no reason to stop at the distribution level reached in generation II: *multi-room* networking (cf. Philips Streamium™). The iPod provides for easy *multi-house* distribution (at friends' houses, from home to office, etc.); legal and standards issues rather than technology issues hamper *multi-premise* sharing of content, with new business models arising, for example, for the proprietors mentioned previously. Computer science research-ers have investigated a next degree, *on-the-move*, where content can be shared even without intervention of the human whose devices "encounter each other" (cf. iClouds; Straub & Heinemann, 2004). The largest, *global* scale has already seen a tremendous boost with the deployment of peer-to-peer (P2P) technology since the advent of Napster; new resource sharing models for (free-rider-preventing) mobile use and the incorporation of FedNodes will be added.

All in one, HE R&D must be carefully lead in order to not restrict itself to multi-room/house scenarios; the wider the distribution range, the more important the consideration of recent computer science research.

Smart Home

It has been argued previously that the pulverization of hardware and software and the amalgamation into application hyper domains go hand in hand. This applies to the different degrees of distribution in ambient entertainment, but it also ap-plies to different goals of smart home technology; again, it is important to draw upon the potential synergies and overcome hurdles on the way to using them and understand complementary research contributions. Quite a number of smart-home-related research projects were terminated without immediate business impact. A dominating reason is the (perceived) unattractive cost-benefit ratio of correspond-ing devices and scenarios for average users. It is therefore crucial to concentrate on—and integrate—areas where this ratio can be trusted to improve the foreseeable future (called first priority areas). In terms of needs and potential cost savings, *as-sisted living* is most promising, especially for the elderly and handicapped. Areas of immediate *saving*, such as fine-grained energy optimization, and areas of high perceived need, as well as *safety/security*, make an ideal complement if re-use can be stressed. At first sight, HE is at the other side of the spectrum: desire rather than need dominates spending. Two arguments support the quest for integrating this smart home subdomain with the others mentioned: (1) the *combined* value of desire- and need-driven consumption justifies prices (and therefore, technology sophistication levels) that would otherwise be inhibitory; and (2) each subdomain contributes to overall solutions with unique strengths; for instance, the user interface (UI) expertise built up in HE is definitely beneficial for areas like safety/security where UIs were underemphasized in the past. Based on the advancements, that is, declining prices to be expected from such a synergy in the first priority areas mentioned previously, second priority areas may finally gain market acceptance. This way, the often dis-cussed smart fridges and smart microwave ovens may finally hit the mass market and communicate with our grocery store—but that remains a long way to go, as opposed to newspaper visions.

Beyond Usability, Broadcast, and TV

Some Reflections on TV Development

Since television emerged as a medium in the early days of the past century, it has been evolving constantly—and so it is today. This process of constant further development does not only concern its external appearance, but is foremost due to the changing of its program structures and offering, its forms of usage by the end user, and thus, how it is perceived as a medium as such in society.

With media digitization and the switch over from analogue to digital broadcasting technologies during the 1990s, the further development of television sped up significantly. Digitalization enabled the merger of formerly separated media domains and technologies of different technology sectors. Commonly referred to as *media convergence* this process can be seen as one of the driving forces of media evolution during the last 15 years.

Media Convergence and its Implications

The concept of media convergence stands for the coalescence and/or mergence of formerly separated media types or media domains such as the IT/computing sector, the broadcasting, and the telecommunications sector. Foremost, three aspects of media convergence have to be considered: (1) the technical convergence of media technologies (from transmission networks to end devices), (2) the convergence of content (program forms, multimedia services), and (3) the functional convergence (convergence of media functionalities and forms of usage). As to the further development of media, all three aspects of media convergence impact each other and go hand in hand with the alteration of the end-user needs and requirements on the different media.

There are three common convergence theories of how convergence happens and influences the current and future development of digital media:

1. **Theory of replacement.** New media will replace old media types and assume their functions.
2. **Theory of media completion.** New media will complement old media in peaceful coexistence and/or enhance it with new functionalities.
3. **Theory of media fusion.** New and old media will merge into "all in one" media, assuming all functions of its predecessors.

Apparently media development proceeds according to a mixture of all three theories of media convergence: On the one hand, history gives evidence that old media will not disappear but instead will advance and mutate constantly, adapting itself to new user requirements. On the other hand, new types of media will not replace former media but instead influence and/or modify them, for example, enhancing them with new functionalities. Beyond, new types of media could also absorb functions of older predecessors, if this means an improvement or a benefit to the end users.

Today, media is diversifying according to altering socioeconomic exigencies and consumer needs: Different complementary media with different program and service offers comply with manifold requirements in order to satisfy specific end-user needs in different usages contexts.

Thus, convergence must be considered as the decoupling and re-modulation of technology, content and functionalities of different media types, enabling on the one hand the further development of traditional media and its enhancement with new functionalities, and on the other hand the creation and establishment of novel forms of hybrid media.

Current and Future Strands of TV Development and its Impact for Broadcasters/Service Producers

The development of digital television today seems to be twofold and along two strands. Driven by the merger of broadcasting and IT/computing technologies, traditional TV becomes a multimedia platform accumulating more and more different functionalities and usage options for its end users. For example, content transmission over heterogeneous networks and additional end-device functionalities also provides end users with the option to use and retrieve TV programs differently—for example, to use content (inter)actively via a backchannel. In return, broadcasters are encouraged to create new multimedia services and novel program forms that correspond to the new requirements of their audience. As a result, interactive television (iTV) and programming established some years ago as one new form of television.

The second trend of the last 10 years is the emergence of several novel hybrid TV derivates. Among them, the most important are Internet protocol television (IPTV) and—as the latest development—mobile TV. Most of such hybrid forms of television provide traditional TV programs—or parts thereof—over other transmission networks for the reception on new-end devices. This allows the TV audience to receive such program offers and services in new usage contexts and to also use it differently according to new functionalities and capabilities of the receiving end device.

Today, the four most popular ways of content and service production for hybrid TV are:

1. The takeover and provision of the complete program 1:1 from traditional

2. To re-purpose parts of the original program or service, enhancing it with additional usage options according to the different functionalities of the end device

3. To combine the original program with content from other media and thus, to create novel program forms and services which correspond to the different

4. The creation of new content and program forms or services, targeted at the specific requirements of the devices and usage context

Thanks to technical innovations such as non-proprietary, scalable, content formats, generic metadata standards and intelligent digital production systems, it will be technically feasible very soon to transmit and display each kind of content on every display for its reception on each end device. This option of complete content syndication is of great benefit for broadcasters and service providers since it opens up for them a "million" new options for service production and provision. Their maximum target is and will remain to be to provide their programs and services to the maximum number of end users on different end devices and usage situations.

However, the duties for broadcasters and service providers in terms of intelligent service and program creation will not become obsolete. Instead, in order to provide services and programs, which offer a real added value to their audience, it will be even more important for them to evaluate beforehand which kind of content really makes sense on which device and in which usage context—and foremost to first consider their end users' interests.

Design Principles for Ambient Interactivity

Within the scope of this section different principles for the creation of ambient interactivity are shown. They shall guide service creators to successfully add interactivity to their applications and services.

Opportunistic Interaction

The introduction and wide adoption of the Web has been promoted and attributed to the interactive nature of the new medium. It often goes without much thought that if something is interactive then it is also preferable. Interactivity with the user might seem as the major benefit of iTV, but this is a fallacy that designers with computer experience should learn to avoid. Most notably, there is evidence that in some cases interactivity may be disruptive to the entertainment experience. Vorderer, Knobloch, and Schramm (2001) found that there are some categories of users who do not like to have the option to change the flow of a TV story; they just prefer to watch passively. Indeed, the passive uses and emotional needs gratified by the broadcast media are desirable (Lee & Lee, 1995). Still, there might be cases such as video games, in which the addition of interactive elements enhances the entertainment experience (Malone, 1982). As a principle, empower the viewer with features borrowed from a TV production studio. For example, users could control the display of sports statistics and play along with the players of quiz games. Interactivity should not be enforced to the users, but should be always pervasive for changing the flow of the running program or augmenting with additional information on demand.

Multiple Levels of Attention

A common fallacy is that TV viewers are always concentrated on the TV content, but there is ample evidence that TV usage takes many forms, as far as the levels of attention of the viewer are concerned. Jenkins (2001) opposes the popular view that iTV will support only the needs of the channel surfers by making an analogy: *"With the rise of printing, intensive reading was theoretically displaced by extensive reading: readers read more books and spent less time on each. But intensive reading never totally vanished."* Lee and Lee (1995) found that there is a wide diversity of attention levels to the television set—from background noise to full concentration. For example, a viewer may sit down and watch a TV program attentively, or leave the TV on as a radio and only watch when something interesting comes up (Clancey, 1994). These findings contrast *"to the image of the highly interactive viewer intently engaged with the television set that is often summoned up in talking about new possibilities"* (Lee & Lee, 1995). Instead of assuming a user, who is eager to navigate through persistent dialog boxes, designers should consider that users do not have to be attentive for the application to proceed.

Content Navigation and Selection

During the 1990s there had been a lot of speculation about the 500 channels future of iTV. In contrast, mass communication researchers found that viewers recall and attend to fewer than a dozen TV channels (Ferguson & Perse, 1993). The fallacy of the 500 channels future was turned upside down into a new fallacy during the next decade, when researchers put forward the vision of a single personalized channel. The study of TV consumption in the home reveals that TV viewing is usually a planned activity, which is a finding that sharply contrasts with the focus on the electronic program guide (EPG) as a method to select a program to watch each time a user opens the TV. Indeed, ritualized TV viewing was confirmed by a survey, in which 63% of the respondents had watched the program before and knew it was going to be on (Lee & Lee, 1995). Still, there is a fraction of the viewers that impulsively select a program to watch, especially among the younger demographic (Gauntlett & Hill, 1999). As a consequence, designers should consider that most TV viewing starts with familiar content, but it might continue with browsing of relevant items. Therefore, iTV applications should support relaxed exploration, instead of information seeking. This principle becomes especially important in the age of hybrid content distribution systems, which include P2P, IPTV, and mobile TV.

Content Delivery Schedule

Using the television as a time tool to structure activities and organize time has been documented at an ethnographic study of a set-top box (STB) trial (O'Brien, Rodden, Rouncefield, & Hughes, 1999). The fact that most TV viewing is considered to be "ritualistic" (Lee & Lee, 1995) does not preclude the exploitation of out-of-band techniques for collecting the content at user's premises. Broadcast distribution is suitable for the delivery of high-demand, high-bit-rate items, which have a real-time appeal (e.g., popular sport events, news). Designers should justify the use of persistent local storage and broadband Internet connections, which are becoming standard, into many iTV products (e.g., video game consoles, digital media players). Digital local storage technology takes viewer control one big step further—from simple channel selection with the remote—by offering the opportunity for conveniently time-shifted local programming and content selection. As a principle, designers should try to release the content from the fixed broadcast schedule and augment it with out-of-band content delivery. Therefore, an appropriate UI for content delivery should allow the user to customize the preferred sources of additional and alternative information and video content.

User-Contributed Content

TV content production has been regarded as a one-way activity that begins with the professional TV producers and editors and ends with post-production at the broadcast station. As a matter of fact, television viewers have long been considered passive receivers of available content, but a new generation of computer literate TV viewers has been accustomed to make and share edits of video content online. The most obvious example of the need for user contributions in available TV content is the activity of TV content forums and related Web sites. There are many types of user communities from the purely instrumental insertion of subtitles in hard-to-find

Japanese anime to the creative competition on scenarios of discontinued favorable TV series. In any case, there are many opportunities for user-contributed content, such as annotations, sharing, and virtual edits. Furthermore, the wide availability of video capture (e.g., in mobile phones, photo cameras) and easy-to-use video editing software, opens up additional opportunities for wider distribution of homemade content (e.g., P2P, portable video players, etc.).

Group Viewing

Just like PC input devices, most TV sets come with one remote control, which excludes the possibility for interactivity to anyone, but the one who keeps the remote control. Despite this shortcoming, TV usage has been always considered a group activity (Gauntlett & Hill, 1999), and it might provide a better experience when watched with family members (Kubey & Csikszentmihalyi, 1990). In contrast, PC usage is mostly solitary, partly because the arrangement of equipment does not provide affordances for group use. Then, a possible pitfall is to consider only one user interacting with the TV, because there is only one remote control. Therefore, designers should consider social viewing that might take place locally. For example, an iTV quiz game might provide opportunities for competition between family members. In the case of distant groups of synchronous viewing, there are further opportunities for group collaboration, which are discussed next.

Content-Enriched Communication

Besides enjoying TV watching together, people enjoy talking about, or referring to TV content (Lee & Lee, 1995). This finding could be regarded as a combination of the previous *group viewing* and *user-contributed content* principles, but in an asynchronous, or distant communication fashion. Therefore, iTV applications should support the communication of groups of people who have watched the same content item, although not at the same time (e.g., family members living in the same or diasporic households). Moreover, iTV applications should facilitate the real-time communication of distant groups of viewers, who watch TV concurrently. An additional aspect of this principle is that it poses an implicit argument against personalization. If TV content is such an important placeholder for discussion, then personalization reduces the chances that any two might have watched the same program. On the other hand, this social aspect of TV viewing might also point towards new directions for personalization, which are based on the behavior of small social circles of affiliated people.

TV Grammar and Aesthetics

A common pitfall, which is facilitated by contemporary authoring tools, is the employment of UI programming toolkits with elements from the PC and the Web, such from buttons, icons, and links (Chorianopoulos & Spinellis, 2004). An additional difficulty in the domain of iTV UI design is the interface's inability to stay attractive over time. TV audiences have become familiar with a visual grammar that requires all programs, as well as presentation styles, to be dynamic and surprising (Meuleman, Heister, Kohar, & Tedd, 1998), which is in sharp contrast with the traditional usability principle of consistency (Nielsen, 1994). In summary, designers should enhance the core and familiar TV notions (e.g., characters, stories) with programmable behaviors (e.g., objects, actions). Then, an iTV UI might not look like a button or a dialog box. Instead, it could be an animated character, which features multimodal behaviors. Furthermore, user selections that activate scene changes should be performed in accordance with the established and familiar TV visual grammar (e.g., dissolves, transitions, fade-outs).

Digital Television, Print Media, and Their Convergence

Increased digitalization since the 1990s caused a transformation in media industry towards an economical concentration and technological integration (media convergence) (Chon, Choi, Barnett, Danowski, & Joo, 2003). Convergence of media aims at increasing the entertainment experience of consumers, which *"appears more and more to be to be a crucial condition for the successful information processing"* (Vorderer, 2001). Especially the future of print media is a challenge as such, due to changing consumer behavior. An increased competition between digital media and print media, forces print media companies to publish their content also on other channels, such as the Internet. And the Internet means an increased active role of the consumer as media aggregator, active user, and a shift of the decision of when and how media are used (Livaditi, Vassilopoulou, Lougos, & Chorianopolos, 2003).

Convergence Between Television and Print Media

Based on the aforementioned empirical findings and the supportive role of entertaining experiences on the users' processing of information we want to point out that the integration of content having print and television origins creates several challenges. If the users consume television and electronic "print" media simultaneously, then we want to focus on three problem areas that may influence their entertainment experience.

First, the content and interaction design needs to address the users' roles of both the print media readers as well as the television viewers. Depending on the genre, the broadcasted print content requires the same quality as its respective printed equivalent. This guarantees that the reader can print out the broadcasted version and perceive no difference between this and a preprinted copy of the same product. Moreover, the navigation and the content interaction require the same simplicity that the user expects from dealing with iTV applications. Furthermore, the provision of print content on the television platform for active media users that access the content at will reveals the problems of the underlying concepts: the push versus the pull model.

Second, the technological provision and distribution of high-quality print content on the digital television platform requires high bandwidths in order to minimize the latency times between the user's executed action to access the print content and its final rendering on the television screen. This applies to both when the content is downloaded to the STB from the object carousel and when it is delivered via other channels such as the Internet. Furthermore, appropriate printing technology needs to be present in order to deliver the user the expected quality of a print product.

Third, the creation, provision, and distribution of the print content require several parties to be involved. A media company creates the print content, provides it digitally, and a broadcast company broadcasts it. Thereby, the print content may be related to the television content. In case the users want to print it out they may delegate the task to an external printing house ensuring the desired quality. Thus, an efficient collaboration between these four parties is necessary to create revenue.

The DigiTVandPrint Project

Regarding the DigiTVandPrint project carried out by the Helsinki University of Technology—VTT Information Technology under participation of several media companies—we wish to explore the provision and distribution of print content over the broadcast network to media consumers using self-developed prototypes. We consider the previous elaborated problem areas as centric. From these perspectives, we state that today's HE systems fulfil the needs for simultaneous consumption of entertaining television and print content only in a limited way. The seamless integration of print content on the digital television platform demands more from the technology. We state that among other factors higher display resolutions, bandwidths, and a complete packet-based television infrastructure are required in order to provide the necessary quality and flexibility in the delivery of the mediated information.

Recent available technological developments such as high definition television (HDTV) and IPTV contribute to this.

Addressing the first problem area, we stimulate the users with interactive HDTV simulations where several experimental variables, such as text length, content genre, or the number of interactive features will be changed in order to find the most crucial factors. We use psychophysiological and subjective evaluation methodologies to research the users' cognitive and emotional states. With regard to information processing, we measure their ability to understand and remember various television and print content. We ask the users to rate their entertainment experience to compare these results with the data gathered during the experiments.

Concerning the second problem area, we plan, design, and implement a prototypical system that allows us to perform tests and thus to find out possible architectural bottlenecks, to develop design guidelines, and to define specific technological requirements. The used technology involves a digital video broadcasting (DVB) test play-out center, various hardware and software components, and digital printers, all partially provided by the named research institutes and the participating media companies.

We develop business models to clarify the technological and economical responsibilities of the involved parties. The results of the research of the first and the second problem area influence these models and deepen the knowledge about possible scenarios and successful applications.

After the first period from May 2004 to May 2005—developing early prototypes and planning initial tests—we are currently planning new experiments and designing prototypical applications that contribute to the pilot tests that are to be performed in spring 2006. We expect to present the final results of the project in November 2006.

Conclusion

It is still to be seen how ambient media will be developing in the future. However, this chapter showed advanced application scenarios for the future of media, which are further elaborated in Lugmayr (in press). Other topics highly relevant for the further development of ambient media include:

- The fact, that in the forthcoming years, mobile services platforms will undoubtedly be deployed into beyond-third-generation (B3G) environments
- Social implication and possible future limitations of rapid technology development in the field of mobile ambient intelligent services

- The characteristics and potential of these innovations in the context of ambient information technology tools for wellness services

It is still to be seen how ambient media will develop during the next years. The year 2010 has been chosen by the European Commission as a target when ambient intelligence should be in place. From this year on very concrete applications should be available to the consumer, and their practical viability will be tested.

References

Braun, E., & Mühlhäuser, M. (2005). *Interacting with federated devices.* Paper presented at the 3rd International Conference on Pervasive Computing.

Chon, B. S., Choi, J. H., Barnett, G. A., Danowski, J. A., & Joo, S. (2003). A structural analysis of media convergence: Cross-industry mergers and acquisitions in the information industries. *Journal of Media Economics, 16*(3), 141-157.

Chorianopoulos, K., & Spinellis, D. (2004). User interface development for interactive television: Extending a commercial DTV platform to the virtual channel API. *Computers and Graphics, 28*(2), 157-166.

Clancey, M. (1994). The television audience examined. *Journal of Advertising Research, 34*(4), 2-11.

Ferguson, D. A., & Perse, E. M. (1993). Media and audience influences on channel repertoire. *Journal of Broadcasting and Electronic Media, 37*(1), 31-47.

Gauntlett, D., & Hill, A. (1999). *TV living: Television, culture and everyday life.* Routledge.

ISTAG. (2001, February). *Scenarios for ambient intelligence in 2010—Final Report.*

ISTAG. (2003). *Ambient intelligence: From vision to reality* (draft report). European Union: IST Advisory Group.

Jenkins, H. (2001). *TV tomorrow.* MIT Technology Review.

Kubey, R., & Csikszentmihalyi, M. (1990). *Television and the quality of life: How viewing shapes everyday experiences.* Lawrence Erlbaum.

Lee, B., & Lee, R. S. (1995). How and why people watch TV: Implications for the future of interactive television. *Journal of Advertising Research, 35*(6), 9-18.

Livaditi, J., Vassilopoulou, K., Lougos, C., & Chorianopolos, C. (2003). *Needs and gratifications for interactive TV applications: Implications for designers.* Paper presented at the Proceedings of the 36th Annual Hawaii International Conference on System Sciences (HICSS'03).

Lugmayr, A. (2003). *From ambient multimedia to bio-multimedia (BiMu).* Paper presented at the EUROPRIX Scholars Conference, Tampere, Finland.

Lugmayr, A. (Ed.). (in press). *Ambient media and home entertainment.* Athens, Greece: Tampere University of Technology.

Lugmayr, A., Niiranen, S., & Kalli, S. (2004). *Digital interactive TV and metadata—Future broadcast multimedia.* New York: Springer.

Lugmayr, A., Saarinen, T., & Tournut, J. P. (2006). *The digital aura: Ambient mobile computer systems.* Paper presented at the 14th Euromicro International Conference on Parallel, Distributed, and Network-Based Processing (PDP 2006).

Malone, T. W. (1982). *Heuristics for designing enjoyable user interfaces: Lessons from computer games.* Paper presented at the 1982 conference on Human Factors in Computing Systems.

Merriam-Webster Online Dictionary. (n.d.). *Media.* Retrieved from http://www.m-w.com/cgi-bin/dictionary

Meuleman, P., Heister, A., Kohar, H., & Tedd, D. (1998). *Double agent—Presentation and filtering agents for a digital television recording system.* Paper presented at the CHI 98 conference summary on Human Factors in Computing Systems.

Nielsen, J. (1994). *Usability engineering.* San Francisco: Morgan Kaufmann.

O'Brien, J., Rodden, T., Rouncefield, M., & Hughes, J. (1999). At home with the technology: An ethnographic study of a set-top-box trial. *ACM Transactions on Computer-Human Interaction (TOCHI), 6*(3), 282-308.

Pering, T., & al., e. (2005). *Face-to-face media sharing using wireless mobile devices.* Paper presented at the 7th International Symposium on Multimedia, Irvine, CA.

Straub, T., & Heinemann, A. (2004). *An anonymous bonus point system for mobile commerce based on word-of-mouth recommendation.* Paper presented at the Proceedings of the 2004 ACM Symposium on Applied Computing.

Vorderer, P. (2001). It's all entertainment—Sure. But what exactly is entertainment? Communication research, media psychology, and the explanation of entertainment experiences. *Poetics, 29*, 247-261.

Vorderer, P., Knobloch, S., & Schramm, H. (2001). Does entertainment suffer from interactivity? The impact of watching an interactive TV movie on viewers' experience of entertainment. *Media Psychology, 3*(4), 343-363.

Section II

Interaction Design

Chapter VIII

The Use of 'Stalking Horse' Evaluation Prototypes for Probing DTV Accessibility Requirements

Mark V. Springett, Middlesex University, UK

Richard N. Griffiths, University of Brighton, UK

Abstract

This chapter describes a technique that utilises established usability evaluation techniques to discover a range of accessibility requirements for digital TV (DTV) viewers with low vision. A study was reported in which two "stalking horse" prototype conditions were tried by subjects performing interactive tasks. These prototypes were not developed technologies but Wizard-of-Oz style conditions. In one condition subjects were asked to use gestures to interact with DTV services, with the screen responding to their hand movements. The other condition used a static keyboard display placed on the table in front of them. Their role was both to probe the efficacy of these approaches and to prompt rich information relating to the subjects abilities, lifestyles, and strategies for interaction. The reported study analyses four viewers with differing types of sight impairment. The reported study was successful in yielding both general concerns about current approaches to DTV display and interactivity design as well as giving significant insights into the possible potential of and difficulties with alternative input methods. The sessions yielded numerous critical incidents, examples of which are reported and analysed. The format also yielded key insights into the way in which individual viewers compensate for diminished vision by using alternative skills such as touch-typing and alternative sensory signals, inductive reasoning and heuristics. The significance of these insights for DTV design and accessibility support is then discussed.

Introduction

Television has thus far been seen as a passive medium in which the user has relatively little control (Artz, 2001). Its role in the lives of individuals has been characterised more as a sit-back entertainment medium rather than a platform for proactive goal-directed action. We acknowledge the likely continued development of distracting "entertainment" applications and recognise that many pioneering examples of interactive television (iTV) services are of this nature. However, the potential of the medium is considerably greater. Indeed, the more established entertainment aspects of the medium may in the future be harnessed with the satisfaction of more substantial long-term goals.

In the short term the potential for maximising access to iTV and TV facilities per se is far from realisation. This is particularly true for certain types of viewers, including viewers who have impaired vision, hearing, or physical manipulation capabilities. Many citizens suffer to a degree from all three conditions, particularly people of advancing age. This makes it difficult to precisely define the condition of a particular viewer. Some recent advances in access research report useful developments including the RNIB Tiresias font (see Gill, 2001), while Pereira (2003) in a comprehensive investigation found that while current DTV services had serious accessibility problems, potential developments such as smart cards had the potential to make a significant difference.

This chapter is partly motivated by the need to tailor set-top box interaction to the needs of individuals in a more informed way. A further motivation is a need for approaches that take a holistic view of user needs rather than simply "designing for a disability." The work builds in part on theory-based descriptions of artefact-based requirements capture described in French and Springett (2003). The approach is designed to use the interactive experience to prompt rich data collection encompassing information about lifestyle, strategy for dealing with visual impairment, and TV preferences along with evaluation data pertaining to specific design prototypes.

A number of known accessibility problems provide context for current investigations. Sight-impaired users report problems searching for mislaid remote control devices, suggesting that either fixed devices or direct communication with the set-top box may be useful alternatives. Conventional remote control design tends to exclude many sight impaired users. Users with hearing and sight impairments have difficulty reading subtitles in their present form. Hearing-impaired users using hearing aids are likely to require a greater level of interaction with TV controls given the variable nature of output and environmental noise. Given these issues the investigation of alternative interaction methods can provide a number of insights. It may be that the choice of input mode is service specific.

Study Design

The aim of the study was to combine the experience of using the set-up "technology" with elicitation of lifestyle-based requirements by indirect means. Therefore, the tasks were presented to the subjects in the form of instructions, in which they were asked to perform specific tasks, namely, finding weather information on the British Broadcasting Corporation (BBC) digital text service and dispatching a mail message via the TV. Both tasks involve a number of atomic actions, such as recognition of both visual and audio information and cursor-based navigation. The first task directed users from the electronic programme guide (EPG) to the News 24 channel, via red button access to the BBC News Multiscreen. The subjects then had to select the appropriate window and appropriate text menu options. This task required them to recognise and manoeuvre the active cursor around a sequential menu (the EPG); scan and select an option; and confirm a selection. The BBC News Multiscreen then displays six labelled options selectable by manoeuvring the cursor (visually cued around the borders of the currently active option) and "pressing" to confirm a selection. The text menu that is then displayed is on a coloured background that is overlaid on the current TV picture. The second task began with activation of the SKY ACTIVE menu followed by a search for the text-messaging option through a number of individual screens, and the composition and despatch of a brief text message. This involved them encountering a number of differently designed screens, cursor designs, and menu displays.

This approach is similar to that described by Ericsson and Simon (1984) in as far as the probe sought to generate critical incident data combined with authentic verbal accounts of the subject's reaction. However, subjects were also encouraged to converse and make any verbal contribution that they considered relevant. This approached was designed to allow the device and the environment to prompt verbalisation of semi-tacit information from subjects.

Method

A simple "Wizard of Oz" protocol was used to animate the prototype iTV interfaces: A secret operator actually controlled the television in response to the manipulations of the prototype by the subject. The studies took place in a usability lab designed to simulate a domestic environment—a lounge with sofas and a coffee table in front of a 26 inch widescreen cathode ray tube (CRT) television. The session was recorded via steerable closed circuit television (CCTV) cameras fixed in the room, the TV display being directly recorded from the set-top box. One wall of the room has a large viewing window, which appears as a mirror from inside the room. The secret operator viewed the trial through this window and operated the TV by a handset extension. The layout is shown in Figure 1.

For the trials involving the mock-up control pad, interpretation of the manipulation was straightforward. Subjects touched the symbols on the pad, and the secret operator pressed the equivalent key on the real controller.

For the gesture driven trial, interpretation was much less objectively precise, as there were no initially agreed semantics for the gestures employed and determining the beginning and end of a particular gesture is not straightforward. However, the simulation was surprisingly effective, with subjects reporting that they believed they really were directly controlling the TV. This was helped considerably by the secret operator having knowledge of the tasks the subjects were carrying out and the subjects' comments as they attempted them. Automating this in a reliable product would not be trivial!

Subjects

Four subjects were invited to take part. Each invitee had a different type of sight impairment.

Subject 1 had Usher type 2 including Retinitis Pigmentosa. Usher Syndrome is a genetic disorder with hearing loss and Retinitis Pigmentosa, is an eye disease that affects a person's vision and peripheral vision; a genetic disorder that is ususally hereditary. Symptoms start with decreased night vision and later progress to the diminishing of peripheral vision. The rate of decline varies depending on the genetic makeup of the disorder and also varies somewhat in individuals' vision (A to Z of Deafblindness, n.d.).

Subject 2 had Macular degeneration. Macular degeneration is the imprecise historical name given to that group of diseases that causes sight-sensing cells in the macular zone of the retina to malfunction or lose function and results in debilitating loss of vital central or detail vision.

Subject 3 had retina problems in one eye and very short sight in the other.

Subject 4 had Leber's optic neuropathy. This condition causes loss of central vision.

The four subjects' profiles demonstrate a range of variants in the type of sight problems that the individuals experienced.

The Prototype Applications

The two input conditions were designed to represent different design criteria. In condition one the subjects were told to use gestures to operate the screen. The rationale behind this design option was that it could remove the need for a secondary remote control device. The subjects will be told that they can gesture at the screen in any way that they regard as appropriate. One of the aims was to investigate whether gestures have potential as a natural form of input.

The second condition involved a mock-up control pad displayed in a fixed position on the table in front of the subjects. The size and layout of the control pad was part modelled on INTELLIKEYS and partly on the SKY remote control layout. Four coloured buttons were designed to correspond to on-screen controls. The rationale behind the choice of a large static keypad was partly to compensate for low vision by enlarging the visual prompts on the device while retaining the visual layout of the original hand-held device. A further consideration was motivated by preliminary interviews with a representative of Sense (a UK based charity representing deafblind citizens) who cited problems with losing ubiquitous remote control devices and conducting lengthy searches to find them.

Figure 2 shows the design of the static keyboard based on the standard SKY remote control device. Figure 3 shows the layout of the control table for experimental condition 2.

Overview Of Results

Three of the subjects were able to complete at least one of the tasks but with considerable difficulty. Subject 1 had difficulty reading the displayed options on the TV and made numerous errors due to misidentifying objects and missing or misinterpreting screen responses. In particular, he was unable to recognise cursor starting positions, or confirmations of selections easily. Subject 2 found that none of the objects words or edges were sufficiently clear to support navigation and was unable to make headway in either of the set tasks. Subject 3 was able to complete the tasks despite repeatedly having to move within inches of the screen, and reporting a difficult and uncomfortable experience. Subject 4 was able to reach the appropriate information in task one. The subject admitted that this was a very difficult task and expressed doubt about whether this would have been possible without prior experience in using the SKY and BBC text services. He was unable to complete the second task even after moving closer to the screen, using it in the manner of a touch screen.

Critical Incident Analysis

We analysed the significant incidents identified in the study using a form of the critical incident analysis. The sessions were analysed to account for the types of statements that were offered by subjects and prompts that elicited that information. A critical incident profile was created for each of the four subjects. The phrase *critical incident* is given a wide-scope interpretation covering significant verbalisations where no current "error phenotype" was cited. The traditional understanding of critical incident tends to refer to occasions during a think-aloud protocol when the subject appears to have difficulty, expresses confusion, or makes mistakes. Subjects are typically encouraged simply to verbalise their attempts to complete the task. The subjects in this study were given the opportunity to "sit-back" and verbalise any related observations or insights if they desired or simply required a break from the (often strenuous) interaction.

The subjects were encouraged to verbalise observations about their lifestyle, suggestions about the design of the set-top box interaction, and so on, rather than simply their attempts to complete the task. Having said this, our assertion is that the activities they were asked to engage in would be critical in prompting and elucidating these insights. Our philosophy was that the subjects had a dual role that could be elicited concurrently. On the one hand they were mainly novice users of the input mechanisms and the packages. The evaluation-style setup therefore could elicit traditional error phenotypes for deeper analysis. On the other hand, the sessions were designed to elicit expert knowledge from the subjects. It is typically the case that visually impaired users will have a greater level of conscious expertise than users without significant impairments. They employ strategies for circumventing the problems caused by their impairment. This can include the use of highly personalised assistive technologies such as eyeglasses. It also includes heuristic strategies designed to compensate for the visual impairment that they have. For example, users with visual impairments may find that they need to develop best-guess strategies for reasoning about the identity of a word or phrase by shape or form. This information is usefully communicated through "demonstration" (in effect talking through an incident in a think-aloud protocol where the strategy is employed). Along with

Figure 1. Laboratory setup for the Wizard-of-Oz evaluation sessions

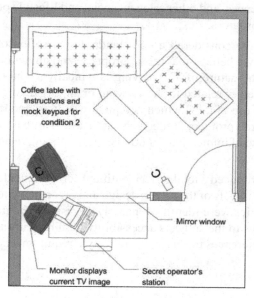

Figure 2. The mock-up static SKY-style remote control, used in conjunction with text and numeric keypads (original dimensions 210mm by 290mm)

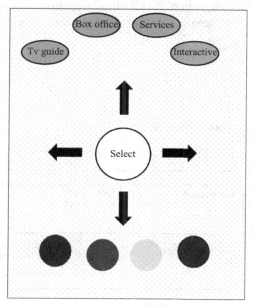

this, the subjects were encouraged to relate their experience of the task to prior experiences, preferences, and other relevant social and lifestyle observations while the session was in progress.

Although this may be considered a non-standard application of the think-aloud protocol approach, we believe that it is appropriate and authentic when applied to TV applications. The nature of TV viewing is typically social or integrated with other activities, whether related to the viewing or simply taking place in parallel. This contrasts with use of a PC which is typically solitary and relatively focused. Therefore the "broken protocol" reflects the authentic nature of TV viewing. This also provides a sense of immersion conducive to stimulating relatively tacit knowledge from the subjects.

The four subjects produced a total of 33 identifiable critical incidents. The majority of these referred to aspects of the screen design and on-screen interactivity support. Three other categories were identified, namely, (1) issues about the input mechanisms; (2) references to the subjects accessibility requirements and strategies; and (3) more general references to service and lifestyle requirements.

Screen Design Problems

The sessions showed that there are numerous problems with the design of displays and support for interactivity where the user has low vision. The session with subject one showed a total of 11 separate critical incidents that were identified as screen-related problems.

Figure 3. The full keyboard layout for condition 2

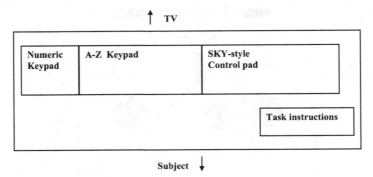

Two recurring problems concerned the overlay of text-based information over an ongoing TV programme. The constant feint movement was found to cause a number of problems both in identification of text items and recognising feedback after input actions.

Recognising options on the SKY EPG was a repeated problem. The display presents pale blue writing on a dark blue background. Subject 1 found that it was impossible to identify complete words while searching the EPG. A further problem was that visual feedback state changes within the EPG were too feint or slight to affect recognition. On numerous occasions an action simply led to a prolonged visual scan to detect any change. Subject 4 reported extreme difficulties in reading text options, particularly where text was overlaid on animated news reports.

The variance of page metaphors within SKY Interactive caused a number of problems. Some of the most significant difficulties were in the identification and manipulation of cursors. The problems fell into three categories, namely, (1) identification of start position; (2) recognition of movement and change; and (3) predictability of action.

The starting position problem occurred on most of the screens visited. Subject 4 reported particular difficulties in this respect. Both subjects 1 and 4 mistook the default position of the cursor for a header line or were simply unable to see it. The cursor in most examples was simply a static image. Viewers need to perform an action to be able to identify it. The cursor does not flash in the fashion of a windows cursor. The subjects therefore found it very difficult to find the cursor. The strategy of finding the cursor on a particular screen through action was not open to subject 1. Therefore he was forced to scan the screen and guess, focussing on candidate objects in turn until finding the active cursor.

Subjects also found that movement and change could be extremely difficult or impossible to detect. As mentioned previously, a number of screens display information overlaid on programmes or trailers. As a result visually impaired viewers have to identify significant movement on the interactive page amid sundry random movements in the background. A further problem was that it was that the physical form of the cursor, and its host objects changed considerably from screen to screen. This made both initial recognition and tracking of the cursor a new challenge as more screens were visited.

A further feedback problem emanated from the "overlay" and the use of parallel scheduled and interactive displays. On one occasion, a selection action coincided with a significant (but unrelated) change in the background video which caused a subject to believe that the last cursor action had effected the change. The subject had appeared to be correctly and very deliberately moving towards successful completion of task two when the incident occurred. The result was that the subject lost faith in their strategy and began speculating about the device's behaviour. The issue of feedback after actions led subject 4 to recommend the use of auditory output to guide navigation.

Predictability of action was also problematic due to difficulty in identifying potentially active objects. For example the EPG has icons at the top of the page that are not active. These were misidentified as active features. On one page there were two separate numbered lists; one selectable the other not. Subject one spent considerable time on this page unable to recognise which parts of the screen were active and which were not. Two pages in particular displayed numbered items that were not actually selectable. This combined with the subjects difficulties in following cursor movement led to significant confusion.

Discussion of Screen Issues

Analysis of the error data suggests that many problems were due either to poor use of colour, movement, and spatial signals, or consistent layout and screen behaviour.

Analysis suggested that many of the identified problems were exacerbated by the lack of an established page metaphor and the consequent effects on learning. The screens visited displayed multiple cursors, contrasting conventions for cursor movement (or selection by keypad in some cases) and a variety of screen layouts. The consequence of this for all user groups is that learning and ease-of-use are impaired significantly. In effect, viewers are required to internalise numerous sets of rules of operation for the various pages. This would make overall competence and skill-based performance with the interactive services significantly harder to attain. The user needs to learn numerous interaction procedures that can be coupled with particular services and recognised on cue.

In the case of visually impaired users this visually led learning process is inevitably slower but is made monumentally harder by the inconsistency of the screens layout and interaction metaphor.

While a number of problems emanate from the high-level problem of inconsistency, there were numerous individual examples in which the size, colour, and general presentation of on-screen objects caused problems. All subjects noted that the colour schemes tended to be inconsistently used, making it hard to recognise objects and distinguish selectable objects. This problem, added to the proliferation of displayed objects, seems to make interaction ponderous and error prone.

Remote Control Devices

The use of gesture was found to have some potential utility. Having said this, the subjects reported some physical discomfort when attempting use gestures continuously over a long period. Two subjects were particularly positive about its potential. The subjects' two-dimensional gestures (up, down, left right) seemed to be reasonably intuitive. Selection or confirmation gestures seemed to pose more problems. Furthermore, the problem of identifying objects on-screen seemed to be the major barrier to its utility. Freehand gesture input offers no extra cue or feedback over and above that which is offered by the screen.

Subject 4 radically adjusted position in order to make out individual EPG options and began touching the screen directly. Subject 1 also knelt close to the screen at points to study on-screen items. Subject 2 could not read the options at all despite using a magnifier. The possibility of using gestures therefore would seem to be limited given that the viewer could not comfortably get close enough to the screen to detect necessary cues. However subject 1 suggested that it could usefully support a limited range of regularly performed actions.

Subjects 1 and 4 had concerns about the layout of the keypad. Subject 4 made the point that the QWERTY keyboard was important for those, who like him, had learned touch-typing in advance of losing central vision. Touch-typing is an example of a compiled cognitive-motor skill. This type of skill requires little in the way of visual cueing, with spatial and sensory information providing the necessary sensory input. Supporting such skilled behaviour allows those with central vision problems to interact fluently rather than painstakingly inspect the remote device for visual information.

By contrast, subject 2 was not a skilled touch-typist and reported considerable problems trying to read the keyboard close up.

Subjects 3 and 4 suggested that speech output could have a profound effect on their ability to interact with DTV. For example, subject 4 suggested using speech output reading the menu options in turn. This would allow immediate feedback which would facilitate both ease-of-use but also learning of spatial information that could support the development of skill-based interaction for frequently visited pages.

Subject 4 reported limited use of DTV for entertainment purposes such as football scores and recommended the wider deployment of magnification facilities that allow the viewer to expand options to a readable size. Subject 1 also made this recommendation, with particular reference to the EPG.

Subject 1 made reference to the use of tactile information, where a remote control device is used. Tactile feedback, such as a clicking sensation when an option is pressed, aid viewers with both vision and hearing problems as these channels would more typically provide this instant feedback.

Discussion of Remote Device Issues

The gesture-based condition exposed potential limitations to the potential of the technique, but the spontaneous behaviour of some subjects suggested a possible alternative.

Although the nature of the Wizard-of-Oz testing eliminated some errors that a fully developed system may cause, there was still evidence that its utility may be limited. The secret operator responded both to gestures and simultaneous verbalisations of intentions, saving the possibility that a number of hand or body movements made by subjects could have unintentional effects. However, on occasion the potential problems with gesture input became clear. On a few occasions this involved subjects causing an effect with a hand movement that was not intended. This suggests that gestures may be hazardous and mistake prone if the set-top box is simply sensitive to movement without some form of opening and closure signal that distinguishes a genuine command from an irrelevant movement.

The second issue involved occasions when there was a gulf between the intentions of the subject expressed in the gesture and the behaviour of the cursor. This problem was compounded by the fact that on occasion this led to confusing screen behaviour which was difficult to interpret and select recovery actions for. Subjects' interpretation of visual feedback is reduced to partial guesswork if they have difficulty seeing objects and object movement. The interpretation is naturally linked to their sense of the goal state that they expect to see. If the action is not what is intended users need sufficient information to interpret the current state of interaction, the effect of their last action, and through this an indication of how to rectify the situation. The problem of inconsistent interaction metaphors described earlier compounds the problem. If a gesture has unexpected effects users with low vision have the extra difficulty of interpreting an unexpected change, which may not have been perceived.

The fact that two of the subjects began touching the actual screen suggests potential for screen-based remote control devices that have displays resembling the screen. The user of such a device is likely to enjoy some critical usability advantages. The remote can be moved to the optimal position for visibility and tactile interaction. The knowledge that the main screen would correspondingly alter as the remote control interface changes would be likely to help alleviate the problem of users with sight impairment not noticing unexpected movements on the screen. Given the variance in the nature and degree of individuals' sight impairments this would afford a degree of personalization. Users would have control over viewing distance, adjusting to their visual capabilities. There is also some reason to think that this would score better than freehand gestures for physical ergonomics. Subjects suggested that extended use of freehand gestures were uncomfortable after a period of continuous interaction. Another issue is that users with sight impairments find themselves having to perform awkward physical contortions or hold uncomfortable positions in order to maximise their ability to see the screen. The ability to move the remote device to various positions again could alleviate such a problem. Furthermore, it has the potential to make the tracking of state changes easier.

The suggestions made by two subjects that gestures could be employed for a very limited range of actions appears to be worthy of further investigation. One problem with gestures were that the screen metaphors made for unpredictable input/output couplings, making it difficult for subjects to recognise and interpret feedback. For gestures to be accessible it seems that the effects must be relatively clear and predictable. Basic TV controls may be a suitable application for gesture-based interaction. Viewers with sensory impairments tend to need such controls relatively often. For example, brightness and contrast may need considerable adjustment depending on the nature of the programme being watched. In the case of Retinitis Pigmentosa, these facilities can address the associated "night vision" issues if the screen is relatively dark. Viewers may find themselves having to make repeated adjustments. The "feedback" would simply be the perceived changes on the screen. While the issue of how to initiate and distinguish these controls is likely to be a continuing issue, this has the advantage of being a simple and consistent interaction. The technique's potential utility for highly interactive and relatively complex applications is highly questionable.

The fixed keyboard condition was praised for the fact that it was stable, but it does not address the physical ergonomic or the perception of feedback issues. There is no guarantee that the user would be able to see labels on the buttons, even though the displayed options were considerably larger than the original SKY remote control. Again, the experimental conditions have to be considered as the remote "display" was on a piece of laminated card, meaning that that the subjects received no tactile information in choosing options or tactile feedback on actions. Two subjects emphasised the advantage of tactile feedback. This feedback includes edges and depth changes when users run their fingers over options. Also, users may be given further feedback if the act of pressing a button produces a decisive feeling of selection, something that the prototype condition could not achieve. The subjects' suggestion that speech output or magnification be used seems to be particularly critical, where there is a highly interactive task to be performed. Typing messages, for example, is prone to input errors, and this is very difficult for those with low vision to monitor.

Further Requirements Information from the Session

While the critical incident analysis was able to highlight important design issues, it also gave a number of insights into the subjects' lifestyle and needs. Our aim was to capture information that informed a range of user-experience aspects, with accessibility issues contextualised in the light of individual subjects. Contributions from subjects fell into two categories, access requirements and lifestyle/service requirements. The former refers specifically to phenomena to do with subject's strategies for interacting with the environment and the implications for access support. The other refers to expressed requirements of a more general nature.

The subjects were able to make rich descriptions of their strategies with minimal prompting, simply adding expanded descriptions of their incident descriptions during interaction. The following segment is a quote from subject 1, describing strategies to compensate for low peripheral vision:

Often you can recognise people by lots of signals like their shape, or how they move, their size, you know, work out whether they're a man or a woman. I can see you from here (3 feet away), but generally that's how I recognise people.

All these signals you get ... you use every resource you have. like if you walk down the street I can see nothing, but when you go from one house to another there's a gap and you can feel the wind coming through so you've an idea of where you are.

This is one of a number of examples in which subjects described ways in which they used other senses and inferential reasoning to compensate for low vision. This has both immediate and long-term significance. One noted problem was the absence of tactile or audible feedback in response to selections and cursor movements. The subjects were only afforded visual cues and feedback, and other senses were wastefully redundant.

Subjects were also on occasion able to make specific suggestions about desirable features. A further example combines a subject's description of their strategy with a specific feature suggestion:

The problem is I haven't got enough sight for those existing sized menus. If there was a facility where I could enlarge any menu I was on, like the Teletext page I could read it. If it was double the size I could read them. You see, how I recognised Scotland I'd recognised the letter s and thought 'what would an 8 letter word with s be, if it were 5 I'd think Sudan or south.' On the Piccadilly tube if you see 2 words, one of 5 characters one of 4 you know you're at Green Park. If it's one word with a V it's Victoria.

The subject is reduced to (somewhat hazardous) guesswork because of the lack of explicit visible information. Thus a rationale for providing a magnification facility is explicated.

A further significant point seems to be that designers have an opportunity to support recognition by form and shape where objects are displayed. Where a page is frequently used, the form and spatial layout become determinants of feature recognition, rather than users reading feature names (see O'Malley & Draper, 1992). In other words, users of a device tend to compile procedures that are stimulated by relatively minimal spatial information, relying on the consistent behaviour of the device. This suggests that the proliferation of page and interaction metaphors is a significant barrier to accessibility for those with low vision.

The information gained about strategy from the subject made reference to lifestyle as a whole as well as technology use. Subject 2 for example reported issues in using customised speech-based accessibility aids, when viewing in social company, suggesting the need for sound-based devices that connect to a remote hearing device.

A further user experience aspect reported on was the threshold of acceptability. Viewers may be willing to learn techniques or use ponderous or awkward devices if they feel that the effort-reward ratio is acceptable. The study revealed that interaction may be possible, but the difficulties render the experience uncomfortable and not enjoyable. Some TV services are social in nature, some are exceptional events with intrinsic appeal, and viewers would be prepared to put up with a degree of discomfort to enjoy them. It is possible to draw analogies with viewers in general, including those without impairment. A room full of football supporters would be likely to continue to watch the World Cup final, even on a badly flickering screen. By contrast DTV services such as form-filling do not seem to have a similar nature or appeal, and there must be some doubt about potentials take-up of such facilities where interaction presents any level of challenge or burden. Any judgement on the accessibility of services and applications should consider the overall user experience rather than merely facilitating input and output.

Conclusion

While novel accessible technologies are often criticised for "one-size-fits-all" approaches, evaluation exercises using these technologies are a valuable source of requirements gathering. Sight-impaired users are domain experts in a similar way to industrial experts in that the extraction of requirements should involve participative situated and indirect probing (Maiden & Rugg, 1997). The approach we describe combines elements of protocol-style evaluation and scenario-based requirements capture. Because subjects were performing tasks interactively, their verbal and physical behaviour provides evidence of mental model development and critical incidents. The conversational nature of the dialogue with the moderator allowed the subjects to expand on any aspect of the task space or relevant aspects of lifestyle as well as task-support needs specifically. A number of requirements were discovered through observations prompted by events during interaction. These can be seen as examples of use of the artefact-prompting expression of semi-tacit knowledge. In some further examples the subjects' behaviour revealed important information such as the production of an eyeglass by one subject.

The hands-on use of the prototypes provides a way of mapping out the design space by drawing contrasts between needs and the nature of the design. Providing accessibility for all involves knowledge about the cognitive and physical usability of the display and its input modes, but this requires contextualisation and an understanding of aspects of lifestyle and strategies for interacting with the environment. Given that many citizens develop multiple sensory and manipulation problems, the range of appropriate solutions is vast and requires tailoring to individual needs. Much of this knowledge may be hard to capture in direct probes such as face-to-face interviews. The use of evaluation exercises provides the stimulus and the context to probe such knowledge effectively.

References

A to Z of Deafblindness. (n.d.). Retrieved from www.deafblind.com

Artz, J. (1996). Computers and the quality of life: Assessing flow in information systems. *Computers and Society, 26*(3), 7-12.

Ericsson, K. A., & Simon, H. A. (1984). *Protocol analysis: Verbal reports as data.* Cambridge, MA: The MIT Press.

French, T. S., & Springett, M. (2003). Developing novel iTV applications: A user centric view. In J. Masthoff, R. Griffiths, & L. Pemberton (Eds.), *First European Conference on iTV.* Retrieved from http://www.brighton.ac.uk/interactive/euroitv/euroitv03/Proceedings.htm

Gill, J. M. (2001). Inclusive design of interactive television. Retrieved from www.tiresias.org/reports/dtg.htm

Maiden, N. A. M., & Rugg, G. (1996). ACRE: A framework for acquisition of requirements. *Software Engineering Journal, 11*(3), 183-192.

O'Malley, C., & Draper, S. (1992). Are mental models really in the mind? Representation and interaction. In Y. Rogers, A. Rutherford, & P. Bibby (Eds.), *Models in the mind: Perspectives, theory and application* (pp.73-91). London: Academic Press.

Pereira, S. (2003). Interactive digital television for people with low vision. Retrieved from http://www.tiresias.org/itv/itv2.htm

Rice, M., & Fels, D. (2004). *Low vision the visual interface for interactive television.* In J. Masthoff, R. Griffiths, & L. Pemberton (Eds.), *Paper presented at the 2nd European Conference on Interactive Television: Enhancing the Experience, Brighton, UK.*

Chapter IX

An Activity-Oriented Approach to Designing a User Interface for Digital Television

Shang Hwa Hsu, National Chiao Tung University, Taiwan

Ming-Hui Weng, National Chiao Tung University, Taiwan

Cha-Hoang Lee, National Chiao Tung University, Taiwan

Abstract

This chapter proposes an activity-oriented approach to digital television (DTV) user interface design. Our approach addresses DTV usefulness and usability issues and entails two phases. A user activity analysis is conducted in phase one, and activities and their social/cultural context are identified. DTV service functions are then conceived to support user activities and their context. DTV service usefulness can be ensured as a result. The user interface design considers both activity requirements and user requirements such as user's related product experience, mental model, and preferences in phase two. Consequently, DTV usability is achieved. A DTV user interface concept is thus proposed. The interface design concept contains the following design features: activity-oriented user interface flow, remote control for universal access, shallow menu hierarchy, display management, adaptive information presentation, and context sensitive functions. Usability evaluation results indicate that the user interface is easy to use to all participants.

Introduction

DTV has several advantages over conventional analogue television: better picture and sound quality, more channels, interactivity, and accessibility. Many countries plan to switch to DTV within the next 5 to 10 years. To facilitate DTV user adoption, Hsu (2005) conducted a survey exploring factors driving DTV diffusion. Identified factors were: government support, reliable technology, price, DTV service usefulness, and easy-to-use user interface. Service usefulness and easy-to-use user interface were among the most important driving factors determining user adoption.

Identifying useful services and designing an easy-to-use user interface is challenging. First, DTV can be complicated for most users. After switching from analogue to digital, television is no longer just a standalone audio-video device. DTV is becoming central to the digital home by providing information services; controlling home appliances and security devices; supporting communication activities; performing business transactions; storing video images; and so on. Therefore, users' manipulation of numerous service applications and easily accessing what they want is critical. Secondly, nearly everyone may use DTV. However, different users may use DTV services for their own purposes and have different usage patterns. Moreover, users interacting with the system have different experiences and abilities. Therefore, accommodating diversified user needs and skill levels is also important.

The user-centered design (Norman & Draper, 1986) has been a dominant approach to user interface design in meeting usability requirements. The user-centered design approach essentially focuses on and includes the user in the design process, beginning with user identification and user tasks. User's target-market characteristics are identified according to user analysis. These characteristics include: demographic characteristics, knowledge and skills, limitations, attitude, and preferences. Based on these characteristics, product requirements can be determined. Task analysis identifies user needs in task performance. These needs include functions and information as well as control, enabling users to perform tasks. Task analysis results are used to specify functional and use requirements. The user interface considers compatibility between user's information processing limitations and task demands, to ensure usability. Since the user-centered approach recognizes that the design will not be right the first time, it suggests that an iterative design be incorporated into the product development process. By using a product prototype, the design can be refined according to user feedback. The user-centered design approach has been successfully applied to many computer products.

However, the user-centered approach has some drawbacks and limitations. Norman (2005) warned against using it in designing everyday products. He argued that the user-centered approach provides a limited design view. It is suitable for products targeted for a particular market and for specific user-task support, but everyday products are designed for everyone and support a variety of tasks. These tasks are typically coordinated and integrated into higher -level activity units. For everyday product design, Norman proposed an activity-oriented design approach. He asserted that a higher-level activity focus enables designers to take a broader view, yielding an activity supportive design. In a nutshell, the activity-oriented approach focuses on user activity understanding, and its design fits activity requirements.

In line with Norman's (2005) notion of activity-centered design, Kuutti (1996) proposed an activity theory application framework to the human-computer interaction design (HCI). According to activity theory (Engeström, 1987), an activity is the way a subject acts toward an object. An activity may vary as the object changes, and the relationship between the subject and object is tool mediated. Tools enable the subject to transform an object into an outcome. Furthermore, an activity is conducted in an environment that has social and cultural context. Two new relationships (subject-community and community-object) were added to the subject-object model. The community is a shareholder group in a particular activity or those who share the same activity objective. Rules and regulations govern how an activity is carried out in a community and impose environmental influences and conditions on an activity. Furthermore, community members assume different activity roles and responsibilities. These components provide a social contextual framework for HCI design. Therefore, activity theory not only focuses on user activity analysis, but also considers cooperative work and social relationship interactions (Kuutti, 1996). That is, activity theory treats HCI as a complete interaction perspective, not limited to interaction between an individual and a computer.

Although several researchers (e.g., Bødker, 1996; Nardi, 1996) have attempted to apply activity theory to HCI design, there is no formal method for putting activity theory into practice. To resolve this problem, Mwanza (2002) proposed a systematic design framework called activity-oriented design method (AODM), based on the activity triangle model (Engeström, 1987). The method contains eight steps aimed at successively collecting information regarding eight activity components: *Subjects, objects, tool, rules, community, division of labor, transformation process* and *outcomes.* Mwanza successfully applied activity theory to two projects by employing this eight-step model.

DTV is intended for individual use in daily life activities. A life activity can include many others (family members, friends, or others) besides the individual, and may include many social processes. Therefore, DTV can be a social device shared by family and friends alike. DTV design should consider social and cultural contexts. Furthermore, life activities are dynamic. They are composed of task sets that change as conditions change. Thus, the design should reflect possible actions and conditions in which people function (Norman, 2005).

This chapter proposes an activity-oriented approach to DTV user interface design. This approach begins with an understanding of life activities and their requirements in use context. Activity theory is used as a framework for life activity understanding. Life activities are treated as analysis units. Analysis on use context includes interaction among community members underlying the activity, rules governing the activity, and physical, as well as sociocultural aspects of the use environment. From activity analysis, activity requirements are determined and, in turn, functional requirements can be specified. After activity requirements are determined, the approach moves to identifying user requirements such as users' experiences, skills, knowledge, and preferences. User interface design tries to optimize compatibility between an activity and its contextual elements so that it not only meets activity requirements, but also satisfies individual as well as social needs. In so doing, DTV usefulness and usability can be ensured.

The proposed activity-oriented approach consists of four steps:

- **Analyze DTV activity requirements.** People engage in an activity in order to attain an objective. The objective is then transformed into an outcome that motivates people to engage in the activity. Accordingly, the activity can be identified by the motive for using DTV. Since the identified activities may be too numerous, related activities are grouped into activity groups for simplicity.

  The intertwined relationships among activity components in each activity group are analyzed. All members and their roles in the underlying activity are identified. Moreover, rules governing DTV use and the social processes underlying the activity are analyzed. Social and cultural issues related to DTV use are identified. Activity models are then constructed to depict relationships between the eight activity components.

- **Envision services to support each activity group.** By analyzing activity requirements and its social context, appropriate application services and contents are envisioned. The envisioning process is based on activity models constructed in the previous stage. Conceptual scenarios developed describe how the conceived services support user activity. A DTV system conceptual model is then developed describing system services, their attributes, and the relationships between services. The DTV system can then smoothly meet activity requirements and prove useful to users as a result.

- **Establish a use model.** A use model delineates how users will use DTV services to carry out their activity. Use model development begins with detailed activity analysis. An activity has a hierarchical structure containing a coordinated, integrated set of actions. Actions are targeted toward a specific goal and are achieved through operations. Conditions will control what operations will be taken, and conditions and goals together will determine tool use. Based on the activity-actions-operations hierarchy and activity flow pattern, a use model is then developed to describe the mapping between DTV services and activity flow. This mapping is used as a basis to determine user interface structure and components.

- **Understand user requirements.** To achieve usability goal, user interface design must consider user characteristics. User characteristic exploration includes two parts: (1) understanding users' previous experiences, and (2) exploring users' mental models and preferences for interaction methods. Users accumulate previous experience from earlier product interaction. Previous experience contains users' knowledge, skills, mental models, and affective responses to the related products. Leveraging users' previous related product experiences can ease users' new product learning and prevent user frustration. Furthermore, users interact with the system based on their system knowledge that is, their mental models. Designing a user interface compatible with a user's mental model enhances product usability. Interaction method preference affects users' product attitude. User interface design must meet these user requirements to enhance usability and increase user acceptance.

Case Study: A DTV User Interface Design

Several qualitative and quantitative methods employed in this study applied the activity-oriented approach. Qualitative methods include ethnographic research, in-depth interviews, and a focus group. These qualitative methods analyze activities and their social/cultural context and conceive DTV services. Quantitative methods were also conducted, assessing activity importance and collecting user characteristic information. A survey was conducted to meet the purpose. The survey comprised of questions regarding user activities; services and contents needs; user experiences with related products; user expectations of user interface style; and user characteristics such as demographic variables and attitudes toward technology adoption. The one thousand questionnaire-sampling scheme—distributed across Taiwan—proportionately represented the Taiwanese population in terms of age, sex, occupation, and geographical location. Of the returned questionnaires, 960 were valid.

Understanding User Activities and Their Context

Ethnographic research conducted observed current life activity practices. Twelve lead users (Rogers, 1995) participated in the research. Observations focused on activities engaged in; when and where they were engaged in; how the activity was carried out; who took part in the activity and their roles; and what rules governed this activity. Observers tried to piece information together using the activity triangle model (Engeström, 1987). In some cases, observation results were insufficient to provide a whole activity system picture. Thus, in-depth interviews conducted collected supplementary information. Through ethnographic research and in-depth interviews, 34 life activities were identified and their activity models obtained.

These activities then served as a candidate activity list in the survey questionnaire. Survey respondents were asked to rate the importance and frequency of activity performance. Exploratory factor analysis conducted explored the interrelationships of 37 activities. A Kaiser-Meyer-Olkin (KMO) test of sampling adequacy obtained a good value of 0.911, the Bartlett's test of Sphericity was significant ($p < 0.001$), and the matrix determinant was in acceptable range. Factor analysis results determined five activity groups (refer to appendix A): (1) entertainment (9 activities), (2) information (6 activities), (3) education (9 activities), (4) transactions (5 activities), and (5) daily living (5 activities).

Identifying DTV Service Needs to Support each Activity Group

A focus group meeting met to envision potential DTV services and contents to support each activity group. The focus group consisted of two DTV R&D engineers and six lead users with DTV experience in other countries. The meeting began with a briefing on the latest DTV technology and service developments by the two R&D

Table 1. Useful services for each activity group

Entertainment	Information	Education	Transactions	Daily living
Video-on-demand	Meteorology	Employment	Financial services	Appliance control
Music-on-demand	Transportation	e-Library	TV banking	Voting
Games	Travel	Games	TV shopping	e-Health care
Travel	Financial	e-Learning	Reserving	Video phones
Horoscope	Employment	e-Newspaper	Ticket Purchase	Lottery
TV shopping	Horoscope		Lottery	Home security
Ticket Purchase	TV banking		e-Health management	e-Health management
News-on-demand	Movie		Commodity	Commodity
Voting	News-on-demand			
Transportation	Games			

engineers. Next, lead users were given the activity model obtained in the earlier stage and asked to brainstorm potential DTV services that would be useful in supporting each activity group. Thirty-seven candidate DTV services and contents were gathered as a result. After identifying potential services, conceptual scenarios were developed, describing service use in carrying out activities.

These 37 candidate DTV services were then included in the survey questionnaire. Respondents in the survey rated the services/contents in terms of their usefulness in supporting life activities. Useful services for each activity group are shown in Table 1. Some services/contents were considered useful in supporting more than one activity group.

Developing a DTV use Model

The second ethnographic research conducted collected detailed activity analysis information to determine identified service and content functional requirements in each activity group. Five families were recruited for the ethnographic research. These families were lead users of 3C products. The research took place on one weekday and one weekend. Observations were made to learn how life activities are performed in the family context and how each family member took part in the activities. Observation results gathered decomposed activities into action and operation levels. Moreover, in-depth interviews requested interviewees to envision how they would use DTV services and contents to support their activities. Use models were then developed based on interview results. These use models served as a basis for determining user interface components and its structure.

Ethnographic research results indicated that each activity group had its own specific use model. For example, entertainment activities are performed in a sequential manner. After users find the right program, they will watch the program for a while before they move onto the next activity. Therefore, a sequential menu structure is appropriate to support this type of activity. On the contrary, an information-searching activity is an iterative process. Users navigate through information space in order to seek for related information. Users may easily get lost and find it difficult to return to their original location. Therefore, a network structure is suitable for this type of activity.

Analyzing User Requirements

Understanding User Experience

Survey respondents rated past experience with related products in terms of ease of use and frequency of use. The ease-of-use score reflects product usability and the frequency-of-use score indicates activity support usefulness. Combining the two scores yielded a good experience index for each product. Results showed that nine multimedia products, four information products, and two communication products were rated as "good experience products." Among these products, respondents with high interaction skill rated television sets, TV games, and personal computer as the top three "good experience products," whereas television sets and TV games were regarded as "good experience products" by respondents with low interaction skill.

Identifying UI Design Features

In-depth interviews conducted identified design features that contribute to good experience. Eight persons participated in the in-depth interview sessions, including two businessmen, two children, two housewives, and two older adults. They were frequent users of "good experience products." In the interviews, participants were asked to describe their interaction experience with TV sets, TV games, and personal computers and then list design features they considered good.

Good design features for TV sets were: (1) control function visibility—all control functions have dedicated physical keys easily located on the remote control; (2) shallow menu hierarchy—most TV menu options are organized in one or two levels. Too many menu layers confuse and disorient users so that they have difficulty reaching the target option; (3) feedback information visibility— TV provides immediate selection and system response feedback. Selection feedback shows immediate user selection. System response feedback provides the function whether the user activates it or not; and (4) automatic setting and display management provision—TV users do not need to memorize and learn how to go back to the initial display management setting. When the user turns off the TV set, it will automatically return to the initial setting when it is turned on again.

TV games are popular entertainment for children. Good design features indicated by the interviewees include: (1) comprehensible icons—users can easily identify menu option icons in games; (2) unambiguous function grouping—games always use a meaningful group function rule so users can easily locate the function; (3) legible text display—TV games fit information presentation into the display space so children can read the instruction; (4) easy-to-use function keys—all function keys are unambiguously mapped to the player's task; (5) context-sensitive help—when playing a TV game, the player does not need to refer to the manual. Help information for each possible problem is always provided to the player throughout the game play.

PC users pointed out several good design features of a PC: (1) interactivity—PC provides interactive functions and dynamic feedback allowing users to use appropriate commands to achieve their task goal; (2) graphical user interface—PC contains easy-to-understand icons and intuitive metaphor that makes users feel they are directly manipulating the device; (3) direct linkage to related information—the hyperlink mechanism helps users easily and efficiently locate information they are interested in; (4) multi-tasking—users can perform more then one task simultaneously. PC users wished that DTV would also allow users multiple viewing of TV programs and services at the same time.

These good design features provide insight into DTV design. They can also be used as design references to ensure interoperability of networked products.

Exploring Users' Mental Models and Preferences

Respondents with low interaction skills, such as housewives and older adults, stated their preference for TV interaction style. In fact, even respondents with high interaction skill expected DTV to retain a TV context for entertainment activities, while good PC design features provide support for information-related activities. TV design features all respondents would like to retain on DTV were: (1) TV screen as a default screen, (2) remote control as the main interaction device, (3) shallow menu structure, (4) automatic display settings, and (5) direct access function keys.

The User Interface Design Concept

This study developed a user interface concept based on both activity and user requirements. The concept and the features are described next.

Activity-Oriented User Interface Flow

The use model obtained from ethnographic research served as the design basis for the user interface structure. In this design, services are provided in accordance with activity needs. In addition, services are embedded in the activity so that users need only focus on the activity without contemplating how to apply appropriate services to support the activity. For example, when the user chooses the activity he/she wants to engage in, he/she may get related services in the service menu. Also, when the user uses a service to perform an activity, the system will automatically display related functions. In so doing, users' cognitive load can be reduced.

Remote Control for Universal Access

Survey data showed that most respondents preferred to use a remote control as the primary interaction device. The remote control design is intended to support users with different interaction skills. Thus, in our design, the remote control is divided

Figure 1. The remote control

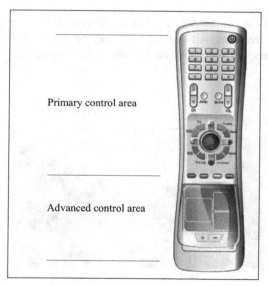

Figure 2. Eight hot keys for activity support

into two areas: the primary control area and the advanced control area (Figure 1). The primary control area is similar to the traditional analog TV remote control. It provides basic function keys (e.g., number key, volume up/down, channel up/down, mute, and so on) sufficient for operating multimedia devices. The primary control area also provides eight function keys dedicated to frequently performed activities identified from the activity analysis. The function keys allow users direct access to critical services or programs for those activities. The primary control area is designed mainly for supporting entertainment and daily life activities. The advanced control area is a touch pad serving several functions: (1) points and selects menu items, (2) inputs text, (3) displays personalized Electronic Program Guide, and (4) manages displays. It is suitable for information, transaction, and education activities.

Shallow Menu Hierarchy

To provide a simple use environment, the menu function items were arranged based on importance and frequency-of-use for each service. The sequential menus were limited to three layers, helping users avoid unnecessary menu traversal. For simpler

Figure 3a. Single-View mode

Figure 3b. Multiple-view mode

operation, menu items can be selected using arrow keys to move the cursor and then pressing the enter (ok) key. Moreover, eight hot keys on the remote control provide direct access to critical activity group services (Figure 2) so that menu traversal is minimized.

Multiple Viewing and Display Management

To accommodate single and multiple viewing behaviors, two display modes (single viewport and multiple viewport) are provided. The single-view mode is designated as a default mode. When the DTV is turned on, it displays single-view mode (Figure 3a). In the single-view mode, if another service or program is selected, the DTV will automatically switch to multiple-view mode (Figure 3b). In the design, the system supports up to four active viewports at the same time. With multiple-view mode display, the user can resize and relocate the viewports at will, or go back to the initial single-view mode by pressing the cancel key on the remote control.

Adaptive Information Presentation

In the multiple-view mode, display information legibility may suffer. To solve this problem, our study developed an adaptive information presentation method. The method contains two mechanisms: menu legibility adjustment and content legibility. Menu legibility affects how many menu options can be displayed. The adaptive information presentation method automatically adjusts the numbers of legible options for viewport size to ensure menu option legibility. Content legibility affects how much information can be displayed. The adaptive information presentation method decides which information is displayed according to its importance. When a display space is reduced, only important information will be displayed, and font size is automatically adjusted to keep information legible (Figure 3b and 4). In addition to the adaptive information presentation method, the system also provides a virtual magnifying glass that enlarges focused content while it shrinks the neighborhood area using a fisheye display method.

Context-Sensitive Menu Functions

To avoid unnecessary menu traversals, functions appropriate to the current use context are provided to users. The context-sensitive menu functions can detect a user's location in the activity menu context and provide suitable activity support options. The functions are displayed in a menu bar and numbered in terms of their appropriateness. (Figure 4)

Figure 4. Context-sensitive menu function

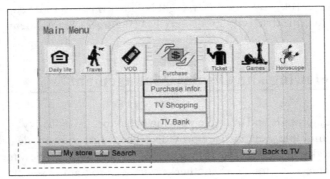

Usability Evaluation

The purpose of usability evaluation is twofold: (1) assessing user interface design usability, and (2) evaluating usefulness of each user interface design feature from the user's point of view.

Participants

Twenty subjects participated in the usability evaluation. In order to represent user population, the recruited participants included 11 high skill level subjects (i.e., three high school students, four college students, four middle-aged technical professionals) and nine low skill level subjects (i.e., three children in elementary school, three middle-aged housewives, and three older adults). They all regarded themselves as early adopters of DTV when they answered the survey question.

Material, User Tasks and Procedures

A user interface concept prototype was developed using Microsoft Visual C++. The prototype simulated the user interface to support three kinds of activities: entertainment, transaction, and information browsing. A liquid-crystal display (LCD) TV was used to display the prototype. The experimental environment is shown in Figure 5.

Before the usability evaluation session began, the experimenter introduced the evaluation procedure and tasks to the participants. In the evaluation stage, subjects were asked to perform tasks by following a test scenario. The test scenario represented

Figure 5. The experimental environment

activities frequently engaged in by family members in a morning. It consisted of three scenarios: (1) searching for news/movie program, (2) browsing information, and (3) shopping on TV. Each activity scenario in turn contained several tasks (Table 2). Specifically, the searching for news/movie program scenario involved three tasks: (1) selecting a TV program from the program guide, (2) browsing a movie from a video-on-demand (VoD) service, and (3) scheduling a program recording. The information-browsing scenario illustrated daily information-related activities such as checking weather, stock information, and traffic information. Finally, the TV shopping task simulated TV-shopping behavior including searching for products, browsing product information, comparing different products, and engaging in online transactions.

When performing these tasks, subjects were asked to think aloud through every task step. After performing the tasks, performance data (task error rate and verbal protocol) were recorded. Upon each task completion, subjects rated the perceived usefulness (five-point Likert scale: 1-non-useful, 5-very useful) and perceived usability (five-point Likert scale: 1-unusable, 5-easy to use) of each user interface feature.

Results

Task Performance

Scenario 1: Searching for News/Movie Programs

In the searching for news/movie programs scenario, almost all operations were error free, except that a few subjects made an error at the step "press function key (*error rate = 0.4*)" in the scheduling a program recording task (Table 3). In order to understand why subjects made such an error, we had a debriefing interview with subjects who made the error. We found that the low skill group subjects expected

Table 2. User's activities, tasks, and operations

Tasks	Sub Tasks	Operations & [DTV design features in the task]
Searching for a News/Movie program	selecting a TV program	(1) power on; (2) key in target program number; (3) press enter
	browsing a movie from a video-on-demand	(1) power on; (2) main menu [shallow menu structure]; (3) select VOD; (4) select program type; (5) select program; (6) start movie
	scheduling a program recording	(1) power on; (2) key in target program number; (3) press enter; (4) press function key [context sensitive functions]; (5) select function; (6) start recording
Browsing Information	checking weather	(1) power on; (2) main menu [shallow menu structure]; (3) select information services; (4) select the item 'weathers' (In TV viewing state, use multiple display): (1) press short-cut bar; (2) select items – 'weather information' adaptive information presentation], multiple viewing & display management]
	stock information	(1) power on; (2) main menu [shallow menu structure], (3) select information services, (4) select the item 'stock'; (5) select a target stock. (In TV viewing state, use multiple display): (1) press short cut bar; (2) select items 'my stock' [adaptive information presentation], multiple viewing & display management]
	traffic information	(1) Power on, main menu [shallow menu structure]; (2) select information services, (3) select the item 'traffic' (in TV viewing state, use multiple display): (1) press short cut bar; (2) select the item 'traffic information'; adaptive information presentation], multiple viewing & display management]
Shopping on TV	TV - shopping	(1) power on; (2) main menu; (3) select purchase; (4) TV-shopping; (5) select shopping store; (6) select target products; (7) compare products; (8) buy the product. [shallow menu structure], [activity oriented UI flow]

Table 3. Error rate of select news/movie programs tasks

Tasks	Sub Tasks	Operations	Error rate
Searching for News/ Movie programs	selecting a TV program	(1) power on	0.05
		(2) key in target program number	0.00
		(3) press enter	0.00
	browsing a movie from video-on-demand	(1) main menu	0.00
		(2) select 'Video On Demand'	0.00
		(3) select program type	0.05
		(4) select program	0.00
		(6) start movie	0.00
	scheduling a program recording	(1) key in target program number	0.00
		(2) press enter	0.00
		(3) press function key	<u>0.40</u>
		(4) select function	0.00
		(5) start recording	0.00

to find the "record" key on the remote control, while the high skill group subjects anticipated that the "record" function was in the main menu on the screen. These expectations were not consistent with our design. In our design, the "record" function is only embedded in the context sensitive function bar. The result caused us to reconsider adding an array of the hard function keys on the remote control for those frequently used functions (e.g., record, play, stop, forward, back).

Scenario 2: Browsing Information

There is only one operational error found in this scenario (Table 4). Subjects made errors in the "select the information item" step of the information-browsing task (error rate = 0.15). After completing the task, we interviewed the subjects and found that three low-skilled subjects did not understand what "submenu" means, because they did not have any computer experience. In revising the menu design, we considered providing a clue to guide the low-skill user group when they interact with the TV menu. Or, we would also provide a number key input for the low skill group to allow them direct function access.

Table 4. Error rate of browsing information task

Tasks	Sub Tasks	Operations	rror rat
browsing information	Access services	(1) power on	0.00
		(2) main menu [shallow menu structure]	0.00
		(3) select information services	0.00
		(4) select the information item	0.15
		(In TV viewing state, use multiple display):	
		(1) press short-cut bar	0.00
		(2) select services items	0.00

Table 5. Error rate of TV shopping task

Tasks	Sub Tasks	Operations	Error rate
Shopping on TV	TV – shopping	(1) power on	0.00
		(2) main menu	0.05
		(3) select purchase	0.05
		(4) TV-shopping	0.00
		(5) select shopping store	0.00
		(6) select target products	0.00
		(7) buy the product	0.00

Scenario 3: Shopping on TV

All subjects not only understood how to perform the TV shopping task but were also able to complete it within a short time. In the debriefing interview, we asked subjects why they were so skilled. Subjects indicated that they felt the TV shopping functions were similar to real life shopping situations. In addition, the function arrangement in the menu structure was easy to understand and navigate.

Evaluation of Usefulness for Each User Interface Design Feature
Activity-Oriented User Interface Flow

All subjects were able to follow the user interface structure to perform TV viewing, information retrieving, and transaction activities with minimal errors (mean error rate = 0.03%). Subjects felt that the activity-oriented user interface flow was useful (mean of perceived usefulness = 4.50, Std. = 0.83) and usable (perceived ease of use = 4.30, Std. = 0.98) for the TV-shopping task. With this design feature, they did not have to search for the service they needed at every task step. That is, the system automatically provided relative information and functions to support the TV-shopping task without traversing through multiple service menus. Therefore, they were not afraid that they might get lost while performing the task.

Shallow Menu Structure

Twenty users gave a high rating of perceived usefulness (mean = 3.60, Std. = 1.23) and perceived usability (mean = 3.80, Std. = 1.11) to this design feature when they completed three information retrieving, VoD, and TV-shopping activities. Subjects were able to complete tasks without making any errors (mean error rate = 0.00). Time recordings of menu navigation indicated that the subjects did not need much time thinking about where to go next. Furthermore, the verbal protocol did not show that subjects experienced any frustration or confusion.

Multiple Viewing and Viewport Management

The multiple-viewing feature was evaluated by the information browsing activity. Technical professionals and students exhibited higher acceptance (i.e., mean of perceived usefulness = 4.07 Std. = 0.37; mean of perceived ease of use = 4.36, Std. = 0.74) of the multiple-viewing mode, while housewives (mean of perceived usefulness = 2.33, Std. = 0.58) and older adults (mean of perceived usefulness = 1.67, Std. = 0.58) had problems with this feature. The former expressed that the multiple-view mode was convenient for them because they could use the information service while they were watching a TV program. However, the latter just wanted to use single-view mode. If multiple-view mode were the only choice, housewives and older adults hoped that the system manufacturer could provide an automatic display management function. Although this design feature may not be well received by housewives and older adults, their task performance was still above the baseline (mean of perceived ease of use = 3.50, Std. = 0.58).

Adaptive Information Presentation

This feature was well received by all subjects (mean of perceived usefulness = 3.85, Std. = 1.27) in information-browsing tasks, especially for older adults. It can enhance displayed information legibility. However, a concern was raised that adaptive information presentation might cause a side effect, that is, increasing font size may decrease the number of menu options displayed and consequently hamper menu interaction performance. However, usability evaluation results indicated that this was not the case. Results revealed that it did not have a negative effect on menu selection or searching for menu options (mean of perceived easy of use = 3.85 Std. = 1.27). In fact, it enhanced menu interaction performance for older adults.

Context-Sensitive Functions

All subjects gave high marks to this design feature. Students and housewives expressed that this design feature was helpful (mean of perceived usefulness = 3.80, Std. = 1.01) and usable (mean of perceived easy to use = 4.00, Std. = 1.08) because it could alleviate efforts to traverse menus and reduce operation time.

Conclusion

This study attempts to apply an activity-oriented design approach to digital TV. This approach analyzes activities and identifies requirements for each type of activity. Activity-oriented user interface flow can help users complete their activities and reduce their workload. The evaluation results support this notion, and services are considered to be accessible to users and useful in supporting activities.

This study also explores good design features of related products used by users for the same activities. Incorporating good design features of related products into the new design not only eases the learning process for the new product, but also helps establish a familiar and comfortable feeling for first-time users. This helps users gain out-of-the-box experience (Ketola, 2005).

We also found that different user groups have different user interface expectations. For example, technicians and students prefer multiple viewing and multiple tasks. On the contrary, older adults prefer a simple TV environment. Therefore, the proposed design concept is flexible enough to accommodate these two types of interaction methods in order to meet two separate user requirements.

Acknowledgment

This research is supported by MediaTek Research Center, National Chiao Tung University, Taiwan.

References

Bødker, S. (1996). Applying activity theory to video analysis: How to make sense of video data in human-computer interaction. In B. A. Nardi (Ed.), *Context and consciousness: Activity theory and human-computer interaction* (pp. 147-174). Cambridge, MA: The MIT Press.

Engeström, Y. (1987). *Learning by expanding: An activity-theoretical approach to developmental research*. Helsinki, Finland: Orienta-Konsultit.

Hsu, S. H. (2005). *Identify user requirements for user interfaces in digital television environment*. Unpublished technical report, Human Factors Laboratory, National Chiao Tung university, Taiwan.

Kaptelinin, V. (1996). Computer-mediated activity: Functional organs in social and developmental con-texts. In B. A. Nardi (Ed.), *Context and consciousness: Activity theory and human-computer interaction* (pp. 103-116). Cambridge, MA: The MIT Press.

Kaufman, L., & Rousueeuw, P. J. (1990). *Finding groups in data: An introduction to cluster analysis*. John Wiley & Sons.

Ketola, P. (2005). Special issue on out-of-the-box experience and consumer devices. *Personal and Ubiquitous Computing, 9,* 187-190.

Kuutti, K. (1996). Activity theory as a potential framework for human-computer interaction research. In B. A. Nardi (Ed.), *Context and consciousness: Activity theory and human-computer interaction* (pp. 17-44). Cambridge, MA: The MIT Press.

Mwanza, D. (2002). *Towards an activity-oriented design method for HCI research and practice.* Unpublished doctoral dissertation, The Open University, UK.

Nardi, B. A. (Ed.). (1996a). *Context and consciousness: Activity theory and human-computer interaction.* Cambridge, MA: The MIT Press.

Nardi, B. A. (1996b). Some reflections on the application of activity theory. In B. A. Nardi (Ed.), *Context and consciousness: Activity theory and human-computer interaction* (pp. 235-246). Cambridge, MA: The MIT Press.

Norman, D. A. (1988). *The psychology of everyday things.* New York: Basic Books.

Norman, D. (2005). Human-centered design considered harmful. *ACM Interactions, 12*(4), 14-19.

Rogers, E. (1995). *Diffusions of innovations* (4th ed.). New York: The Free Press.

Appendix: Factor Loading of Each Activity Group

Activity Group (5 factors)	Information	Education	Entertainment	Daily living	Transaction
Internet	.788	.191	.173	.243	.413
Watching movie	.704	.454	.385	.064	.488
Newspaper	.673	.200	.270	.309	.335
Bookstore shopping	.616	.257	.203	.193	.268
Shopping	.502	.175	.289	.142	.029
Magazine	.440	.228	.301	-.147	.368
Information showing	.372	.665	.500	.015	.475
Social service	.127	.659	.451	.062	.310
Art exhibition	.470	.611	.490	-.007	.483
Accomplishment course	.294	.594	.242	.262	.398
School	.239	.587	.370	.267	.252
Refresher course	.312	.565	.294	.153	.378
Photography	.395	.547	.489	.210	.414
Tai-Chi chuan	-.108	.471	.250	.178	.187
Handicraft making	.214	.445	.361	.263	.106
Traveling	.422	.462	.611	.063	.397
Joy riding	.413	.311	.559	.325	.341
Ball gaming	.002	.450	.535	.011	.272
Body-building	.297	.357	.503	.082	.476
Dinner party	.259	.241	.492	.035	.173
Camping	.355	.437	.481	.446	.364
Ball game watching	.136	.250	.478	.308	.283
Singing (Karaoke)	.236	.338	.476	.145	.185
Religious activity	.009	.375	.433	.008	.127
Housekeeping	.305	.170	.104	.661	.243
Computer gaming	.362	.145	.082	.615	.285
Pet breeding	-.035	.356	.377	.527	.312
Gardening	.021	.370	.262	.441	.400
Cooking	.217	.328	.279	.412	.129
Commodity buying	.346	.341	.158	.360	.670
TV / e-shopping	.300	.301	.338	.384	.638
Investment	.300	.527	.343	.303	.538
Finical news	.444	.360	.407	.227	.498
Lottery	.194	.387	.375	.001	.460

Chapter X

Long-Term Interaction:
Supporting Long-Term Relationship in the iTV Domain

Christoph Haffner, University of Kiel, Germany

Thorsten Völkel, University of Kiel, Germany

Abstract

This chapter introduces the application of concepts for long-term interaction to support long-term relationship in the interactive television (iTV) domain. While classical interaction concepts cover short-term interaction cycles only, theoretical models for long-term interaction and relationships deal with time periods exceeding the human short-term working memory. The user must be supported by memory cues to resume interrupted long-term interactions immediately. The iTV domain offers many long-term interaction scenarios in the context of establishing long-term relationships of recipients and broadcasters. The authors adopt concepts for long-term interaction towards iTV and develop a basic classification of long-term interaction. Three scenarios within the iTV domain illustrate the potential impact for the design of iTV applications.

Introduction

When interacting with electronic systems, users build a mental model regarding what can be done with the system and how it performs its functionality. Consequently, many people have an intuitive understanding of the capabilities offered by devices used in their day-to-day life, though this understanding does not necessarily need to match the actual functionality offered.

Based on the mental model, actions are formulated and executed within an execution-evaluation cycle described in detail in Norman (1986). Figure 1 presents a graphical representation of Norman's model. The cycle consists of three major parts, namely *goals, execution,* and *evaluation,* which are carried out in the cited order when performing a task. A goal is defined in respect to what the user wants to happen, while the execution is how the user actually interacts with the device or system (regarding the input level). Within the evaluation phase, the user compares the final outcome to the intended result of his/her actions.

In detail, seven stages of the model can be identified. Within the first stage, the user forms a goal which might not yet be related to the system. Within the second stage, the user forms an intention including specific actions necessary to reach the goal. These actions are specified within a task sequence that may be related to the actual device. Finally, the fourth stage consists of the physical action thus carrying out the necessary actions with the device. During the three stages of evaluation, the user perceives the state of the device as the result of his/her actions. The results are interpreted and this interpretation is then compared to the intended goals. If the perceived outcome does not match the intended goal, the cycle will begin again with modified actions which the user pretends to produce the desired results.

As the users' mental model does not fully correspond to the actual functionalities of the device or system, difficulties during the interaction are likely to occur. Norman (1986) refers to these problems as the *Gulf of Execution* and the *Gulf of Evaluation.* The first refers to the difference between the possible functionalities of the system and the users' intentions. The second refers to the users' difficulties in interpreting the system's physical state and outcome thus interpreting whether the desired result has been reached.

Norman's (1986) model assumes that the outcome of the actions is available immediately and hence the user is able to correlate them with the causal actions. This assumption is certainly true for short-term interaction. In contrast, long-term interaction spans a temporal interval for the execution of actions and its result longer than short-term memory captures memes. As a consequence, the evaluation part becomes far more difficult for the user, thus the Gulf of Execution significantly increases. In Dix, Ramduny, and Wilkinson (1998) this problem is referred to as the "broken loop of interaction." Cues are needed for the user to regain context when a specific interaction is resumed.

Figure 1. Scheme of Norman's model of interaction

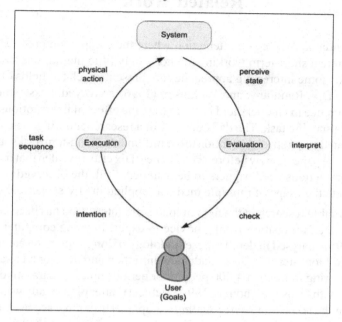

Concepts within the field of long-term human-computer relationship cover the evolution of the engagement between the user and a computer system (or a specific software) over a longer period of time. Recent research results such as Bickmore and Picard (2005) show that strong long-term relationships between human and computer can be built and maintained by designing appropriate long-term interactions.

The convergence of long-term interaction and long-term relationship in the field of iTV applications seems a promising approach for a stronger recipient retention because the long-term relationship between the recipient and the broadcast station or a specific programme is of great interest.

In this paper we discuss the benefits of well-designed, long-term interaction techniques for building and maintaining strong relationships within the iTV domain. Examples for long-term interactions include classic electronic program guide (EPG) applications as well as online quizzes in sequels, betting, and online gambling, or applications to manage private stock portfolios.

The paper is structured as follows, in section 2 we give a brief overview of related work concerning long-term interaction and relationships in human-computer interaction (HCI). In section 3 we introduce a classification of long-term interactions as a basis for the development of iTV scenarios in section 4.

Related Work

As most research is focused on interaction where the completion process does not exceed the human short-term working memory, only little literature is available in this area. Still, some interesting approaches are present in related fields of interest. For example, Dix, Ramduny, and Wilkinson (1996) conveyed a case study at the HCI'95 conference in Huddersfield investigating the effect of interruptions towards carrying out a specific task. The daily activity of a researcher during the conference was investigated given special attention to the fulfilment of long-term tasks such as the procedure of the paper review before a conference. Dix et al. revealed that interrupted long-term interactions need triggers to be resumed. Still, the observed procedures did not include the usage of multiple media as applied in iTV scenarios.

Most considerable research in the field of long-term interaction has been conducted in exploring user's transition from a novice to expert in using computer systems. Thomas (1998) analysed in depth the methodology of longitudinal research in HCI. He assorted all long-term studies conducted until then and developed a large scale survey monitoring more than 4,000 people for periods up to 7 years, subdivided in cohorts each with 50 users (Thomas, 1998). Subject matter of the study was the long-term usage of the text editor *sam* providing a graphical user interface by considering previous word-processing experience. To collect information on the long-term usability issues of the editor, events were logged over the total period of usage (Cook, Kay, Ryan, & Thomas, 1995). Main goals of the Sidney survey were to understand the development of methods of interaction with the editor based on a large sample. Data collection should be adequate, whereas monitoring had to be conducted at a low-cost level while recognising the user's privacy when obtaining data.

The different surveys figured out that experts in using a complex software develop only after a few years of practise (at least about 3 years) and only few people adopt the best methods for solving a task (Thomas, 1998). One of the Sidney study results was the idea of changing the user interface design during the learning process over a long time period, such as an evolutionary user interface attending the user on his/her way from novice to expert.

In Bickmore and Picard (2005) techniques for constructing, maintaining, and evaluating long-term human-computer relationships are investigated. A relational agent was developed who adapted during the time of a long-term interaction in contrast to classical agents like the well-known Microsoft "Clipit" agent, which does not consider previous interactions of the user. They define HCIs as relationships which must also offer, besides reliability and consistency, a certain kind of variability for keeping the user engaged within the interaction. Although the authors offer a comprehensive introduction in the field of long-term human-computer relationships, broadcasting specific aspects imposed by the requirements of iTV applications are not covered.

Classification of Long-Term Interaction

Based on the few surveys and definitions for longitudinal interaction cycles it is not yet possible to develop a general model for long-term interactions. However in a first step it is useful to elaborate a basic classification of different types of long-term interaction in terms of implementing concrete approaches. Before exploring some scenarios more deeply, in the following section we introduce basic classes of long-term interactions being important for the following discussion. We distinguish three main classes of long-term interaction relying on Dix et al.'s assumptions of an interrupted interaction cycle. Namely, we distinguish *externally interrupted long-term interactions* (EI), *nested interrupted long-term interactions* (NI), and *time interrupted long-term interactions* (TI).

Interactions of class EI are separated into smaller parts of interaction. The separation is due to an external event or an externally triggered interaction, thus the interaction may be interrupted within a critical phase. Consequently, the user must spend much effort to regain the context after the interruption. Consider Figure 2 illustrating an example for an externally interrupted interaction.

The user needs to switch to a new context to execute the interrupting interaction 2. Finally, the user must regain the context of the primary interaction or, more precisely, of the exact point where the user stopped the execution. As context switching is very time consuming and for many users a straining task, cues are needed to facilitate the resumption of interaction 1.

Although interactions of class *II*, NI, are very similar to the first class, the event or interaction that triggers the interrupt is closely related to the primary interaction and must be finished before continuing with the interrupted interaction. Consequently, the context of the interrupting interaction is more closely related to the original one.

Figure 2. Externally interrupted long-term interaction

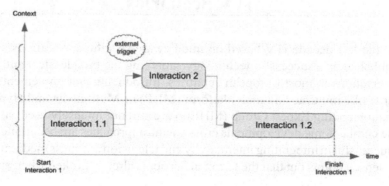

Figure 3. Time interrupted long-term interaction

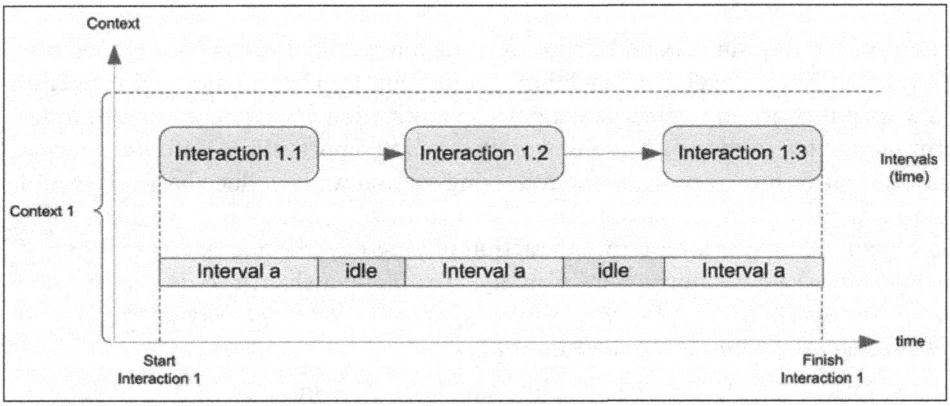

The interruption will normally take place at a point where the original interaction can easily be resumed. Context switching is therefore easier than in the first case.

For class *TI*, the interaction is again separated into parts. The separation takes place directly after the completion of one part of the interaction and is not the effect of an external event or interaction. This class contains interactions which can be easily separated into parts. These parts can then be performed at different points in time. The time between the parts might span some hours or even days. Again, to regain context after a long period, cues are helpful to facilitate the resumption of interaction for the user. Figure 3 illustrates the TI.

After the introduction of some basic classes of long-term interaction, related scenarios are presented in the next section.

iTV Scenarios

During the last decade iTV based on middleware-specific applications could not be established as a successful technology satisfying the broadcasters and customers expectations in most European countries. Introducing and implementing open standards like multimedia home platform (MHP) or Multimedia and Hypermedia information coding Expert Group (MHEG) are still not intensely used. In contrast to these cumbersome developments cross-media approaches are becoming increasingly successful in integrating interactivity and television as a mass medium (Jensen, 2005). Jensen points out that the terminal barrier is already broken when applying a cross-media solution. This keeps costs for producers and customers low and allows fast development of applications.

In most cases television and short message service (SMS) are combined to implement the return channel. Favoured applications are voting, games, and communication formats like chat. Due to response time intervals of less than 1 second, controlling action games via SMS is possible and commercially operated. Application and interaction design of current iTV applications focuses almost on short-term interaction cycles, even if there are underlying long scale time intervals like in EGPs or enhancements of TV series.

While investigating applicable long-term interaction scenarios, we took the cross media as a criteria as well as the basic condition of a time interval exceeding human short-term working memory. Furthermore, we considered that most people prefer interactive formats specifically designed for television. Here one has to follow the low-attention-span paradigm assuming television viewers to intend for short attention time and quick decisions, so called "lazy interactivity" (Jensen, 2005). Confronting these criteria, the most challenging task is the integration of an appropriate level of complexity of long-term tasks with the lazy interactivity.

Beyond the different approaches of implementing interaction, the term *long-term relationship* is of increasing interest in connection with interactive media and the design of user interfaces. Long-term relationship in combination interactivity has a potential to support the assumption that interactivity significantly intensifies the relationship between broadcasters, advertisers, and the viewer. Bickmore and Picard (2005) emphasised the relevance of personal relations to execute tasks in interactive media. The following socio-psychological models were used to develop long-term relationships between human and computer:

- Emotional support
- Appraisal support
- Instrumental support

Concrete applications were implemented as relational agents, visualised as avatars with human appearance.

Emotional support covers, for example, esteem reassurance of worth, affection, trustworthiness, or intimacy. Appraisal support affects significantly the user's orientation within the interaction process. It provides advice and guidance as well as information, cues, and feedback. Instrumental support gives the material assistance that supports the user in handling devices to interact with the application. In addition to those different support types, group belonging and social network support can be of substantial interest in a process of building communities around specific programmes such as series or sports.

Models describing long-term interaction in context of iTV differ from those defined for classical desktop systems as iTV provides the framework for applying the three classes of long-term interaction described previously.

Since Norman's (1986) interactive cycle applies to a short time period, most use cases focus the usage of one single medium, for example, a personal computer. Long-term interaction a priori considers the usage of different media and devices within one single interaction cycle. Winogard (2002) extends existing interaction models by defining interaction spaces within which he dissolves the association of interaction and single devices. This allows a distributed interaction model. In this context he develops an interaction structure that is many to many. One individual can use different devices that can again be used by several different individuals. This model can be applied to todays and future utilisation of iTV. On the one hand TV devices are mostly used by one ore more individuals simultaneously, while mobile television on small devices (DVB-H or DMB-based) will be widely used individually. The consequence is a divergence of programme, device, and location.

The scenarios discussed next shall illustrate the possible implications of long-term interaction concepts and their role in establishing long-term relationships in the TV domain.

Creating Personalised EPGs

EPGs allow the viewer to get an overview on all programmes offered by the current network. The programme listings are usually organised by channel, time, and genre or by actors. Jensen (2005) listed some features describing advanced EPGs such as search engines supporting interactive searches, reminder functionality informing the viewer on favourite programmes, automatic recording of previously selected programmes, personalisation, and customisation as well as intelligent agents checking for certain types of programmes and remembering viewer's preferences.

Being a significant added value for customers, EPGs have to fulfil several requirements (Quico, Costa, & Damásio, 2005). Implementations of long-term interaction within the process of composing an individual TV programme are usually future-oriented activities. The recipient defines an individual play list that reminds him of forthcoming programmes or changes in the programme at specific points in time. Advanced models describe personalised programme guides implementing recommendation systems. In order to specify the viewer's preferences, Adrissono et al. (2004) define a hybrid user model, differentiated in three main areas:

- **Explicit user model** (information the system got from the user)
- **Stereotypical user model** (predictions based on prior user preferences)
- **Dynamic user model** (derived from user behaviour observation).

These models illustrate the long-term relationship between user and EPG application and emphasise the necessity of models for long-term interaction in order to support the long-term relationship sustained.

If the point in time defined by the user is not immediately before the programme and the recipient's current location is unsure, or he/she does not use TV, the reminder is sent as SMS or e-mail. The reminder offers the option to continue the reminding process or to stop it. If the reminding process is continued, the recipient will receive messages until a final reminder is triggered directly on TV. All reminders contain the necessary cues for resuming the interaction. The programme title, airtime, and history of interaction steps help the recipient to continue the interaction immediately. The reminding process may last some weeks or days in advance, considering a maximum time interval that covers the time the programmes are scheduled in advance. Exceptions within the scheduling process are either unanticipated changes in the programme or the case that the air time of a certain movie is not yet defined.

Online Betting in iTV

Online betting and wagering allow gambling via iTV applications. Lotteries or sport wagers can be set and the relating programme shows corresponding events or drawing of lots. Interactive betting in televisions is forecasted as a segment with extensive economic potential in online gambling markets (Haffner, Dahm, & Weimann, 2005).

Interactive applications enable the recipient to set bets on horse races or soccer games before watching them live on TV. When several bets are set on different games or races taking place simultaneously, the recipient is only able to follow one of those at a time. Still, in some cases it is useful for the user to get information about the other games. Obviously, the results of other matches and potential winning notifications are important and can be sent immediately after the game finished. Consequently, the recipient is enabled to continue the interaction cycle by using potential profit and setting new wagers.

To increase the experience of multiple bets, intermediate results of all games can be displayed. For example, if the one team scores a goal, the result and the potential impact upon the specific wager will be displayed on the screen. The interruptions during one interaction are thus increasing the excitement of the user. Additionally, the notification of bet results can be transmitted over several channels like SMS or e-mail. Each notification contains the history of all steps necessary to set the online bet to allow the recipient as easy as possible to continue or finish the interaction cycle. The notification trigger is given by the time when the bet served its purpose.

One potential cross-media application is the building of communities around on-line betting. By introducing forums people are able to discuss bets and wagers and potential chances of winning. Additionally, statistics about the amount of money that has been bet on a team or game can be provided. These statistics might give the user an indication about the general mood concerning the specific wager. Further-more, as is often the case for lottery games, users could place a bet together which will strengthen the community and long-term relationship regarding the gambling application.

Television Sequel

TV sequels are mostly broadcasted as daily or weekly programmes. In general, one season covers a couple of episodes over several months. The concept of a TV sequels aims to establish a long-term relationship between the protagonists and the audience based on a rising pretended intimacy. In combination with the recipient's curiosity on the story's progress, this emotional aspect can be supported by interac-tive enhancements.

In the given scenario a quiz game spanning the whole season is developed. Parts of the quiz are presented to the recipient during each episode. The quiz runs towards the end of an episode and asks for contents of the actual programme and for possible future developments in the story.

The recipient is then informed about results at the beginning of each episode that affect the running total of kudos on his/her account. Having a minimum of kudos at the end of the season, he/she participates in a lottery game.

A variation of this scenario is the convergence of iTV and a Web application. During one episode a protagonist plays a computer game that is to be replayed by the recipient

Figure 4. Nested interrupted long-term interaction in a sequel context

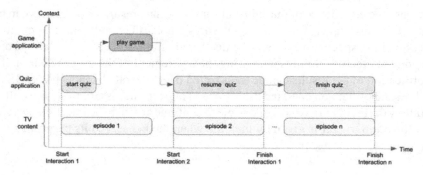

via the Internet. To build a deeper context to the sequel, the recipient receives hints needed for advancing within the quiz game. This nested and interrupted long-term interaction is illustrated in Figure 4. When the recipient resumes the quiz within an episode, the application presents a summary of all preceded steps and the current score of answers he/she gave.

Recipients who join the programme later in time have to reach the current state that other recipients already have. As a concrete example, the recipient has missed the first 3 episodes out of 10 and wants to join the iTV application. All broadcasted episodes are stored within an archive that can be accessed via the Internet. A new recipient can consequently start watching the missed episodes on the World Wide Web. Additionally, he/she is able to play the quiz and game up to the last broadcasted episode.

A future variety of cross-media enhancements of TV sequels could be the interactive drama. This concept incorporates the possibility to influence the strand of the plot (Szilas, 2005), which is basically interesting for TV sequels which are broadcasted over a long period of time.

Custody Account for Private Investors

Since stock dealing is getting a more significant factor in private asset formation and retirement provision, it is an appropriate scenario for long-term interaction in a domain dealing with a vast amount of information within mid- and long-scale time intervals.

In television, a number of programmes deal with financial and stock exchange topics. In iTV the combination of actual information and applications for managing private stock portfolios is a self-evident use case (see for example the Bloomberg eTV service on Sky Digital in UK).

This scenario describes a system supporting customers in managing their stock portfolio with an iTV application supported by an e-mail-driven reminding and recommendation system. It provides another example for an EI illustrated in Figure 5.

While most owners of a private stock portfolio are not professionals in that area, it is the assumption that an easy-to-use system has to consider that the portfolio is increasing over several years in value and amount of stocks. When not dealing in everyday business with stock exchange information, the portfolio owner wants to have appraisal support for understanding the decisions for buying and selling over the past years. The system relies on the assumption that the portfolio owner does not have all information on the stock holding process of all his/her stocks at any time. This basic knowledge includes:

Figure 5. Externally interrupted long-term interactions in stock portfolio management context

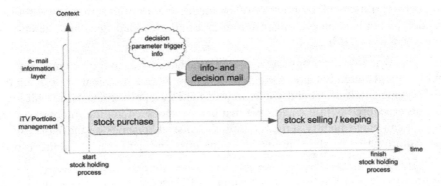

- Reasons for purchasing a stock at a certain point in time (stock price, company data, development information etc.)
- Personal annotations (short summary for reasons of purchase)
- Definition of parameters for information trigger (e.g., stock price—including other stocks)

Within the iTV application, the portfolio owner always has access to all information about his/her portfolio including the parameters for triggering an e-mail that recommends the selling of stocks or only informs about automatically triggered selling processes and the result.

This e-mail includes all information necessary for the portfolio owner to comprehend his/her decision. Since the user relies on that information to decide about buying or selling stocks, the presentation design of the information becomes essential.

Scenario Discussion

All four scenarios illustrate potential situations where recipients are involved in interaction cycles over a long period of time. Periods for long-term interaction vary from several hours, days, weeks, to even months.

While personalisation of EPGs is a subject of interest in HCI (Smith & Cotter, 2005), long-term interaction is not discussed explicitly. The previous EPG example illustrates the situations in which long-term interaction cycles are relevant. In combination with personalisation concepts, long-term interaction has the potential to set up a long-term relationship between a broadcaster or EPG provider and the recipient.

The EPG scenario allows EI to leave the control on the recipient's side.

For interactive, value-added services in iTV, long-term concepts can help to implement easy-to-use applications. In online gambling scenarios, contextual cues must assure the immediate identification of the correct context. The long-term aspect of these cues is emphasised by the custody account scenario. The cues affect the decision-making process of buying and selling stocks within a trustworthily environment. In this scenario, the expertise of the information and TV provider is combined with the users own knowledge and appraisal, leading to a specific decision.

The TV sequel example underlines the opportunities causing associations between TV content and application content aiming to intensify one-way emotional relationships between recipient and protagonists. The sequel as a specific type of programme defines the time periods the recipient deals with (daily, weekly, monthly). As a consequence, the scenario defines the interaction cycle as well and allows NI. The scenario emphasises a notable potential to affiliate convergence of long-term interaction and relationship in iTV to establish a stronger recipient's retention.

Beyond the discussed scenarios, there is a wide variety of additional use cases for long-term interactions, where long-term relationship is aspired in a commercial environment. Examples are interactive advertising or TV shopping programmes. Interactive advertising already tries to attract viewers from the continuous broadcast stream into an interactive application like gaming. Adding a long run application could be an appropriate method for establishing customer retention. Other scenarios could be play-along quizzes offering the opportunity to go through several turns on different days.

The discussion clarifies the fact that there are many examples in the domain of iTV exceeding human short-term memory where the concepts for a systematic implementation of long-term interaction are applicable.

Conclusion and Future Work

In this chapter we present a basic discussion of the importance of the concept of long-term interaction concerning the establishment and maintenance of long-term relationship within the iTV domain while supporting interaction in multiple contexts. A basic classification of long-term interaction is introduced and related to iTV scenarios. Many scenarios of everyday business already deal with middle- and long-term time intervals, but a general model from which applications can be inferred is still missing.

Well-designed, long-term interactions including the usage of multiple media are becoming increasingly important to maintain a strong relationship to the television viewer. In particular iTV applications are appropriate for the realisation of the concepts introduced. Still, future research is needed to investigate user behaviour when an interaction is performed over a long period of time. Especially questions about the type of cues needed for the user to regain context are of great interest. Additionally, within the iTV domain only few real-world applications exist that follow concepts of long-term interaction. Yet, the convergence of the two concepts of long-term interaction and relationship is not adequately treated for iTV applications. Currently, we are designing a general architecture to describe long-term interactions within an abstract layer. Hence, we are going to implement an iTV application to conduct an empirical study to support our theoretical approaches.

References

Ardissono, L., Gena, C., Torasso, P., Bellifemine, F., Difino, A., & Negro, B. (2004). User modeling and recommendation techniques for personalized electronic program guides. In L. Ardissono, A. Kobsa, & M. Maybury (Eds.), *Personalized digital television. Targeting programs to individual viewers* (pp. 3-26). Dordrecht: Kluwer Academic Publishers.

Bickmore, T. W., & Picard, R. W. (2005). Establishing and maintaining long-term human-computer relationships. *ACM Transactions on Computer-Human Interaction, 12*(2), 293-327.

Cook, R., Kay, J., Ryan, G., & Thomas, R. C. (1995). A toolkit for appraising the long-term usability of a text editor. *Software Quality Journal, 4*(2), 131-154.

Dix, A., Ramduny, D., & Wilkinson, J. (1996). Long-term interaction: Learning the 4 Rs. In M. J. Tauber (Ed.), *Conference companion on human factors in computing systems: Common ground* (pp. 169-170). New York: ACM Press.

Dix, A., Ramduny, D., & Wilkinson, J. (1998). Interaction in the large. *Interaction With Computers, Special Issue on Temporal Aspects of Usability, 11*(1), pp. 9-32. Amsterdam: Elsevier.

Haffner, C., Dahm, A., & Weimann, K. (2005). An interactive MHP service for a horse race betting platform. In J. F. Jensen (Ed.), *Proceedings of the 3rd European Conference on Interactive Television. User Centred ITV Systems, Programmes and Applications* (pp. 105-110). Aalborg: InDiMedia.

Jensen, Jens F. (2005). Interactive television: New genres, new format, new content. In *Proceedings of the 2nd Australasian conference on Interactive entertainment* (pp. 89-96). Sydney.

Norman, D. A. (1986). New views of information processing: Implications for intelligent decision support systems. In E. Hollingel et al. (Eds.), *Intelligent decision support in process environment* (pp. 123-136). New York: Springer.

Quico, C., Costa, & Damásio, M. J. (2005). User centered design methodologies applied to the development of an electronic program guide: The partnership experience of PT multimédia and Universidade Lusófona. In J. F. Jensen (Ed.), *Proceedings of the 3ʳᵈ European Conference on Interactive Television. User Centred ITV Systems, Programmes and Applications* (pp. 43-49). Aalborg: InDiMedia.

Smyth, B., & Cotter, P. (2004). Case-studies on the evolution of the personalized electronic program guide. In L. Ardissono, A. Kobsa, & M. Maybury (Eds.), *Personalized digital television. Targeting programs to individual viewers* (pp. 53-71). Dordrecht: Kluwer Academic Publishers.

Szilas, N. (2005). The future of interactive drama. In *Proceedings of the 2ⁿᵈ Australasian conference on Interactive entertainment* (pp. 193-199). Sydney.

Thomas, R. C. (1998). *Long term human-computer interaction. An exploratory perspective.* Berlin/Heidelberg/New York: Springer.

Winogard, T. (2002). Interaction spaces for twenty-first-century computing. In J. M. Carroll (Ed.), *Human computer interaction in the new millenium* (pp. 259-276). New York: ACM Press.

Chapter XI

Proven Interaction Design Solutions for Accessing and Viewing Interactive TV Content Items

Tibor Kunert, Technical University of Ilmenau, Germany

Heidi Krömker, Technical Univeristy of Ilmenau, Germany

Abstract

This chapter describes interaction design solutions for interactive television (iTV) applications. A user task-based approach to interaction design guidance for iTV applications is suggested to easily integrate with a user-centred application development process. Based on the analysis of existing applications, several generic, iTV-specific user tasks are presented. For these user tasks, proven interaction design solutions were empirically identified by usability testing eight broadcasted applications. Specific design solutions to support the user tasks accessing content item and viewing content item are described. Usability test results and design guidance on the design of menus, video multi-screens, indexes, content presentation areas, paging, and scrolling are presented. Target audiences of this chapter are the iTV user interface designers and usability researchers.

Introduction

This chapter describes interaction design solutions for typical user tasks supported by iTV applications. To identify the design solutions, the following steps were conducted:

1. Identification of recurring and content-independent user tasks for iTV applications (generic user tasks, e.g., accessing content item and viewing content item).

2. Competitive usability testing of different interaction design solutions in eight broadcasted iTV applications.

3. Ranking of interaction design solutions that support specific user tasks based on the usability test results.

4. Description of the identified interaction design solutions on two levels:

    a. User interface (UI) elements, for example, menu or video multi-screen, which support specific user tasks.

    b. Design of UI elements, for example, arrow keys or number keys to select menu items.

Background

For iTV applications only very limited interaction design guidance is available. Published iTV-specific design guidance is limited in applicability by a wide range of designers because it is either middleware specific, for example, for the multimedia home platform (MHP) (Rinnetmäki, 2004) or broadcaster specific (British Broadcasting Corporation [BBC], 2006). For the available design guidance that is independent of middleware and broadcaster, unfortunately the method for its development has not been documented leaving it unclear if and how the guidance is based upon empirical evidence (Gawlinski, 2003; Lu, 2005). Only very little iTV-specific design guidance based on empirical evidence has been published (Kunert & Brecht, 2005). International standards on usability of interactive applications, for example, the standard 9241 of the International Organisation for Standardizaton (ISO) (ISO, 1996-1998) and ISO 14915 (ISO, 2002-2003) offer some guidance but lack the consideration of the iTV-specific context of use (Brecht & Kunert, 2004; Brecht, Kunert, & Krömker, 2005). General usability principles also offer some guidance but are too abstract to offer utile and practical support in the application design process (Shneiderman & Plaisant, 2005, pp. 74-82; ISO, 1996; Molich & Nielsen, 1990). Also, they do not consider the TV production process (Krömker & Klimsa, 2005).

In order to support the development of easy-to-use applications interaction design guidance needs to integrate with a user-centred application design and development process. User-centred application development includes the analysis of the user tasks at the beginning of the development process (Mayhew, 1999; Nielsen, 1993; Preece et al., 1994). Especially interaction design decisions are strongly based upon the user tasks that are to be supported by the application under development. While user tasks are "the activities to achieve a goal" (ISO, 2003, p. 2) a goal is "an intented outcome" (ISO, 2003, p. 2). A task analysis "means understanding users' work or play" (Redish & Wixon, 2003, p. 923). However, the step from the goals and user tasks to be supported (*problem space*) to interaction design (*solution space*) is often difficult. The study presented in this chapter aims at bridging the problem space and the solution space to support the user-centred design of easy-to-use iTV applications. After the selection of the goals and user tasks to be supported they can be mapped more easily with interaction design solutions when the design solutions directly refer to specific user tasks.

Generic User Tasks for iTV Applications

The goals to be supported by an application under development often vary from application to application. However, the user tasks supporting the different goals are often the same. They are often simply combined differently for different user goals. Some user tasks are recurring over and over again across different iTV applications. With the aim to discover generic and recurring user tasks for iTV applications, several broadcasted iTV applications were analysed regarding the user tasks they support. As a result 18 generic and content-independent user tasks for iTV applications were extracted and categorised in five user-task categories (Table 1).

Generic user tasks can be combined differently to support different goals. For example, the user goal "Informing about the weather in Brighton tomorrow" consists of the following generic user tasks: 1.1 Becoming aware of available application, 1.2 Starting application, 1.3 Informing about system status, 2.1 Accessing content item, and 2.2 Viewing content item. This set of generic user tasks can be used early in the user-centred application development process to select user tasks for user goals to be supported by the application under development.

Table 1. Generic user tasks for iTV applications and their categories

No.	Generic user task category	No.	Generic user task
1	Using application basics	1.1	Becoming aware of available application
		1.2	Starting application
		1.3	Informing about system status
		1.4	Sizing application
		1.5	Going one level up
		1.6	Exiting application
2	Browsing/searching for content or function	2.1	Acessing content item
		2.2	Viewing content item
		2.3	Sizing video stream
		2.4	Becoming aware of context-sensitive content/function
		2.5	Using help
3	Communicating	3.1	Voting/answering a multiple-choice question
		3.2	Writing/form filling
		3.3	Informing about costs
4	Adapting	4.1	Adapting content
		4.2	Adapting content presentation
		4.3	Adapting content level
5	Learning	5.1	Comparing test answers

Usability Testing of iTV Applications

Objective

The objective of the usability tests was the identification of proven interaction design solutions to support the design of easy-to-use iTV applications. To identify proven solutions different interaction designs were evaluated regarding usability. Usability is empirically evaluated by measures for effectiveness, efficiency, and satisfaction (ISO, 1998). To offer design guidance for iTV applications in general the identified design solutions need to be independent of any specific content, application, or middleware. It is distinguished between two categories of interaction design solutions:

- UI elements, for example, menu or video multi-screen that support specific user tasks

- Design alternatives for UI elements, for example, arrow keys or number keys for selecting menu items

For usability testing of iTV applications it is essential to consider the iTV-specific context of use (Pemberton & Griffiths, 2003). To structure this chapter the Common Industry Format for Usability Test Reports is roughly followed (ISO, 2006).

Tested Applications

We tested eight broadcasted iTV applications (Table 2). The applications were tested from September 1st to October 8th, 2004, indicating the tested version of the applications. The BBC Olympics 2004 application was tested using a prototype instead of the broadcasted application. The prototype was made available by the BBC and is identical to the broadcasted application.

Method

Since no absolute measure of usability exists no interactive application or interaction design solution can be claimed as being useable or easy to use without a point of reference. Therefore, the interaction design solutions were evaluated as proven regarding usability by comparing their usability measures to other design solutions' usability measures. The usability tests conducted for this study were between-subjects competitive tests (Dumas, 2003, p. 1108; Nielsen, 1993, p. 178; Shneiderman & Plaisant, 2005, p. 148). Emphasis was on formative evaluation rather than on sum-

Table 2. The usability tested iTV applications and their content category, broadcaster, and country

Content category	Application	Broadcaster, Country
News	BBCi News	BBC,UK
	Sky News Active	Sky, UK
	Bloomberg Interactive Television	Bloomberg Television, UK
Sports	BBCi Summer Olympics 2004	BBC, UK
Entertainment show/ reality TV	The Farm	Five, UK
	Brainiac	Sky, UK
Music	MTV Interactive	MTV, UK
Morning magazine	This Morning	ITV, UK

mative evaluation, which is aimed at measuring an application's overall usability (Dumas, 2003, p. 1108; Nielsen, 1993, p. 170; Shneiderman & Plaisant, 2005, p. 144). However, the conducted tests were no typical formative usability tests since the motivation was not to optimise the applications' interaction design. Instead, specific UI elements were evaluated and compared with each other to identify proven interaction design solutions.

Test Participants

The selected test participants had previous experience in using interactive media (e.g., Internet, DVD, teletext, mobile phone). Although only 5% of the participants had previous iTV experience, 66% tested more than one application in the test series. Most of the participants were very Internet experienced; average age was 28 years. Most of the participants were employees and some were students of the University of Brighton, UK. To control the participants' similarity that is needed for valid, between-subjects, competitive usability tests, a questionnaire on demographics, media usage, and experience was used. Each application was tested by six participants. By testing with six participants a compromise was made between the reliability of the test results and the costs of conducting usability tests (Nielsen, 1993, p. 173).

Test Tasks

The participants were asked to carry out specific typical user tasks. Each test task represented a user goal consisting of one or more generic user tasks (Table 1). For example, the user goal *"Knowing if the weather tomorrow allows for a walk"* is represented by the following test task: *"You are planning to take a walk on the beach tomorrow. Please look up the weather forecast for Brighton."* It consists of several generic user tasks: "2.1 Accessing content item" and "2.2. Viewing content item" assuming that the application had already been started before. As another example, the test task "There is something happening in the football match you are watching that you want to pay full attention to. Please exit the application." consists of the generic user task "1.6 Exiting application."

Since the tested applications offer only a limited number of functionality most of the applications were fully tested. If participants could not complete a task on their own and asked for help, specified assists were given by the test administrator. To provide consistent assistance across participants for each task, the optimum path for completion was analysed during test preparation. Each task needed to be started from the end of the previous task. To have all participants start each task from a consistent starting point, the participant was guided to the starting point for the next task in case a task could not be completed in the given time.

Test Environment

The applications were tested in the iTV usability lab of the University of Brighton, UK. The tested applications were broadcasted applications running on a Sky DVB-S set-top box and interacted with via the Sky remote control. The BBC Olympics prototype was running on a personal computer. To allow interaction with the prototype using a TV remote control, the remote control signals were mapped to computer keyboard stroke signals using the software Girder 3.2 and the plug-in Igor SFH-56. For the signal transfer an infrared-receiver was connected to the personal computer. Thus the interaction with the prototype was nearly realistic indicated by the participants not noticing that they are in fact interacting with an application running on a personal computer instead of on a set-top box.

Procedure

On arrival participants were informed that the usability of iTV applications was being tested to find out if the applications meet the users' needs. They were told that it was not a test of their abilities. Participants were shown the usability lab, including the control room, and informed that their interaction would be video and audio recorded. They were asked to sign a release form. Participants were asked to verbalise their thoughts, feelings, experiences, expectations, doubts, and so forth, during task completion (thinking-aloud protocol). The test facilitator was sitting next to the participants reading aloud the tasks one after the other and informed the participant when a task was completed. Assistance for task completion was only given after the participant asked for it. The participants had a limited amount of time to complete each task. The time constraint was 5 minutes for each task being the same for all tasks. If the task was not completed within the given time the task was rated as uncompleted and the test proceeded with the next task.

After task performance the participants were interviewed on critical incidents during task completion and on their impressions and opinions. The questionnaire on demographics and media usage and experience was filled out by the participant between task performance and interview. This allowed the test administrator to prepare the interview by consulting with the test assistant regarding the critical incidents during task performance. The interview lasted approximately as long as the task performance and was also recorded. Finally, the participants were given a small present for their participation.

Usability Measures

The selected usability measures refer to the usability metrics effectiveness, efficiency, and satisfaction (ISO, 1998, 2006).

- **Measure for effectiveness:** Effectiveness was measured by the unassisted task completion rate. That is, the percentage of participants who completed and correctly achieved the task goal without assists. An assist is a direct procedural help given by the test administrator to the participants in order to allow the test to proceed in case the participant cannot complete the task.
- **Measure for efficiency:** Efficiency was measured as number and gravity of problems encountered during unassisted task completion. This was accomplished by observing task performance and by interpreting the subjective expressions of the participant during task completion by the thinking-aloud protocol. Time for task completion was not used as an efficiency measure because there is a considerable time delay by the application's loading time depending on a variety of factors, for example, the application's software architecture and the used set-top box. Therefore time on task would not reflect task efficiency.
- **Measure for satisfaction:** Satisfaction was measured by interpreting the subjective expressions of the participants during task completion by the thinking-aloud protocol as well as in the semi-structured interview after task performance.

Method of Usability Test Analysis

Objectives

The test analysis served the identification of proven interaction design solutions to support specific user tasks. It was distinguished between two types of interaction design solutions:

- UI elements to support specific user tasks
- Design alternatives for UI elements

Method to Identify UI Elements to Support Specific User Tasks

The tested applications offered different UI elements to support the same user task. The objective was to identify those UI elements that are the most proven to support a user task. To achieve that objective several steps were carried out:

- It was analysed which UI elements were used by the test participants to complete each user task.
- The usability measures of each user task were compared to each other across applications.
- The UI elements supporting a user task with the best usability measures were considered as the most proven UI elements to support this user task.

Unfortunately, there is neither a standardised set of UI elements for iTV applications nor a standardised set of UI element categories for interactive applications in general (Dumas, 2003). Therefore the identified UI elements were named and defined for the purpose of this study (see the section *Usability Test Results*). For the completion of some generic user tasks more than one UI element was used in the tests. These UI elements all received the same usability measures being the same as the usability measures for the corresponding test task. Also, some UI elements support more than one user task. To keep the task relation of the usability measures for UI elements these UI elements were analysed separately for each task. Therefore one UI element can have more than one usability measure depending on the corresponding task.

Method to Identify Design Alternatives for Specific UI Elements

In the tested applications the same UI element often differs in its design. The objective was to identify the most proven design for each UI element. To achieve that objective several steps were carried out:

- The test measures were assigned to the UI elements supporting the completion of the corresponding task. Thus, usability measures for UI elements related to a specific user task were gained based on their level of task support.
- The usability measures of the same UI element supporting the same user task were compared to each other across different applications.

- The design of a UI element with the best usability measures was considered as the most proven design alternative for this UI element in regard to task support.

The design alternatives for each UI element were described based on their values for specific variables that were identified for the purpose of this study. As an example, the UI element "Menu" can be varied, for example, regarding the following variables: amount of menu entries, position on screen, and navigation within the menu.

Usability Test Results

Ranking of Interaction Design Solutions Based on their Usability Measures

In order to create a ranking of the different interaction design solutions, the measures for effectiveness, efficiency, and satisfaction were combined to a single usability score for each design solution. The three measures were weighted the same. The values for the single usability measures were added up and divided by three resulting in a single percentage number expressing the design solution's level of usability.

Due to space limitation only the results for the generic user tasks „Accessing content item" and „Viewing content item" are presented here as two of the most important user tasks.

Results for "Accessing Content Item"

To support the user task "Accessing content item" the following three UI elements were identified:

- **Video multi-screen/video mosaic:** Presentation of multiple video streams on one page of an iTV application (Table 3).
- **Menu:** Hierarchical structure of navigation menus and submenus in an iTV application (Table 4).
- **Index:** Sorted list of all or of selected topics of an iTV application (Table 5).

Table 3. Examples for the UI element "Video multi-screen." From left to right: BBC Olympics, Sky News Active, BBC News

Table 4. Examples for UI element "Menu"

Table 5. Examples for the UI element "Index." Left: BBC News. Right: Sky News Active

Each of these UI elements supports the generic user task "Accessing content item" differently. This is expressed by their different values for the usability measures. Based on the usability values a ranking was created for "Video multi-screen," "Menu," and "Index" in regard to their level of task support (Table 6).

For each of these UI elements different design alternatives were identified. The tested applications included three different design alternatives of a video multi-screen, eight different menu designs, and two index designs. The different design alternatives differ in regard to specific variables. The following variables were identified:

Video multi-screen:
- Number of presented video streams
- Order of video streams

Menu:
- Amount of menu entries
- Position on screen
- Remote control keys
- On-screen navigation indicators

Index:
- Screen layout
- Index structure

The design alternatives of one UI element are characterised by their different values for these variables. As for the UI elements, each design alternative support the user task "Accessing content item" differently. This is expressed by their different values for the usability measures. Based on the usability values a ranking was created for the different design alternatives of the UI elements in regard to their level of task support (Table 6). Screen shots of all design alternatives for the three UI elements are presented in Tables 3 through 5. The presentation order of the screen shots from left to right represents their ranking.

Results for "Viewing Content Item"

To support the generic user task "Viewing content item" the following UI elements were identified:

- **Content presentation area:** The screen space where an individual content item is presented, for example, consisting of video stream(s), text, and/or still image(s) is presented in Table 7.
- **Paging elements:** If a content item needs more space than is available on the screen, paging elements allow for switching between different application pages. The entire text changes (like in analogue teletext) (Table 8).

Table 6. UI elements to support the generic user task "Accessing content item" with their ranking and usability values (bold and italic numbers). The different design alternatives for each of the UI elements are described by variables and variable values. The different design alternatives for the UI elements are presented with their individual ranking and usability values.

Generic user task	UI element	Ranking			Accessing content item				
			Variables	Variable values (Design alternative)	Ranking	Usability values in %			Implemented in
						Effectiveness	Efficiency	Satisfaction	
	Video multi-screen	*1*				*89*	*97*	*90*	
			Number of video streams/	5/One large video stream with vertical list	1	100	100	100	BBC Olympics (BBC)
			Order of video streams	8/Rectangle	2	100	100	80	Sky News (Sky)
				6/Rectangle	3	66	90	90	BBC News (BBC)

Table 6. continued

		Description		92	86	73	
Menu **2**	Amount of entries/ Position/ Remote control keys/On-screen navigation indicators	3/Middle left/Arrow keys (up/down)/Arrows (up/down) in selected text line	1	100	100	80	This Morning (ITV)
		7/Top right/Arrow keys (up; down)/Arrows (up/down) in selected text line	2	100	80	90	BBC Olympics (BBC)
		4/Middle left/Arrow keys (up/down); no on-screen navigation indicator	2	100	90	80	Brainiac (Sky)
		6/Bottom left/Arrow keys (up/down)/Arrows (up/down) at static position	4	83	90	80	Sky News (Sky)
		7/Bottom left/Arrow keys (up/down); Arrows (up/down) in selected text line	5	100	80	70	BBC News (BBC)
		6/Middle right/Arrow and number keys (up/down); numbers	5	100	90	60	Bloomberg (Bloomberg Television)
		5/Middle/Arrow keys (left/right); no on-screen navigation indicator	7	83	80	70	MTV Interactive (MTV)
		5/Middle left/Arrow keys (up/down); Arrows (up/down) at static position	8	66	80	50	The Farm (Five)
Index **3**				**83**	**70**	**70**	
	Screen layout/ Index structure	Full screen/List	1	83	60	80	BBC News (BBC)
		Full screen with embedded ¼ video stream/Categories	1	83	80	60	Sky News Active (Sky)

Table 7. Examples for the UI element "Content presentation area." Top (from left to right): BBC News, This Morning, Brainiac, Sky News Active. Bottom (from to right): BBC Olympics, MTV, The Farm, Bloomberg

Table 8. Examples for the UI element "Paging elements." From left to right: Bloomberg, BBC News, Brainiac

Table 9. Example for the UI element "Scrolling elements". Left: Sky News Active. Right: Bloomberg

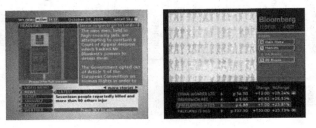

- **Scrolling elements:** If a content item needs more space than available on the screen, scrolling elements allows vertical scrolling. All text lines move up one line, and one text line is exchanged at a time (like in personal computer applications) (Table 9).

Based on the usability values a ranking was created for the UI elements "Paging elements" and "Scrolling elements" in regard to their level of task support. For the UI element "Content presentation area" no ranking and no total usability measures are included because there is no alternative for this UI element (Table 10). However, six different design alternatives were found for it. The tested applications also included three different design alternatives of paging elements but only one design of scrolling elements. Again, specific variables were identified for the design alternatives:

Content presentation area:
*   Screen layout
*   Transparency

Paging elements:
*   Remote control keys
*   On-screen indicators

Scrolling elements:
*   Remote control keys
*   On-screen indicators

As for the UI elements, each design alternative supports the user task "Viewing content item" differently. Again, this is expressed by their different values for the usability measures. Based on the usability values, a ranking was created for the different design alternatives of the UI elements in regard to their level of task support (Table 10). Screen shots of all design alternatives for the three UI elements are presented in Tables 7 through 9. The top-down presentation of the screen shots represents their ranking.

Proven Interaction Design Solutions

In the following, proven design solutions are described that might serve as interaction design guidance for iTV applications.

Table 10. UI elements to support the generic user task "Viewing content item" with their ranking and usability values (bold and italic numbers). The different design alternatives for each of the UI elements are described by variables and their variable values. The different design alternatives for the UI elements are presented with their individual ranking and usability values.

Viewing content item

Generic user task	UI element	Ran-king	Variables	Variable values (Design alternative)	Ran-king	Effectiveness	Efficiency	Satisfaction	Imple-mented in
							Usability values in %		
Content presentation area		-	Screen layout/ Transparency	Full screen with embedded ¼ video stream/20% transparent	1	100	100	100	BBC News (BBC)
				Full screen with embedded ¼ video stream/Non-transparent	2	100	80	80	This Morning (ITV)
				Full screen with embedded ¼ video stream/30% transparent	2	100	80	80	Brainiac (Sky)
				Full screen with embedded ¼ video stream/Non-transparent	2	100	80	80	Sky News (Sky)
				Overlay/Non-transparent	5	100	80	70	BBC Olympics (BBC)
				Overlay/Non-transparent, partly 30% transparent	6	100	60	80	MTV interactive (MTV)
				Overlay/30% transparent	6	100	60	80	The Farm (Five)
				Overlay/Non-transparent	6	100	80	60	Bloomberg (Bloomberg Television)

Table 10. continued

					83	80	80	
Paging elements	**1**	Remote control keys; on-screen indicators	Arrow keys (up/down)/Arrows and position indication (1/3)	1	100	90	90	Bloomberg (Bloomberg Television)
			Arrow keys (up/down)/arrows (up/down) and position indication (Page 1 of 3)	2	84	90	90	BBC News (BBC)
			Colour keys (yellow/ blue)/ coloured dots with text (prev/ next)	3	66	60	60	Brainiac (Sky)
					84	75	60	
Scrolling elements	**2**	Remote control keys; on-screen indicators	Arrow keys (up/down)/arrows (up/down) without position indication	1	100	90	90	Bloomberg (Bloomberg Television)
			Arrow keys (up/down)/arrows (up/down) without position indication	2	84	70	60	Sky News (Sky)

Design Solutions for Accessing Content Item

The three tested UI elements supporting the user to access content items were evaluated as being quite usable as indicated by their single usability scores: Video multi-screen 92%, Menu 83%, and Index 74%. Therefore, they can all be considered as proven design solutions.

- **Design of video multiscreens:** The three tested design alternatives of video multi-screens are all quite proven. The number of presented video streams does not seem to really have an impact on the effectiveness and efficiency since five and eight video streams got an unassisted task completion rate and efficiency value of 100%. However, one big video stream with the others being much smaller seems to be the best solution.

- **Design of menus:** The menu position does not seem to really influence the usability values. Not surprisingly some of the menu designs with only a few items received the highest usability score. However, the menu designs with more items are also quite usable. Either way on-screen navigation indicators seem to be a proven solution regardless of the used remote control keys.

- **Design of indexes:** Index as list and index with categories are both equally usable. More important than the index design seems to be the integration of an index in applications with large amounts of content in general.

The number of applications in which a specific UI element was evaluated indicates the grade of our confidence that this element is in fact proven to support the user task.

Design Solutions for Viewing Content Item

Design of the content presentation area: A full screen layout with embedded ¼ video stream with little or no transparency seems to be the best solution. However, small overlay applications also received good usability values.

Paging Versus Scrolling: Paging is preferred when a content item cannot be displayed on one screen. The up and down arrow keys on the remote control together with graphical arrows (up/down) such as on-screen navigation indicator and position indication (e.g., "Page 1 of 3") have proven most usable. However, it is best when the on-screen arrows are located close to the content items' end to be seen most easily.

Discussion

It seems to be a good idea to design a UI element in the same way as it is designed in the application with the best test results. However, the design solutions identified as proven by this method are easy to use only compared to the design solutions of the other tested applications as point of reference. Other interaction design solutions may exist that simply were not tested.

The identified interaction design solutions are a place to start the UI design of an iTV application. Any UI needs to be built of a combination of elements and needs some internal consistency. This might prevent the iTV designer from simply using the best element for each individual task. He or she needs to find a compromise between consistency and a high level of individual task support. Also, a particular UI element might be good for one task and bad for another when the tasks share one piece of functionality. The iTV designer should make his or her decision based on the priorities of the different user tasks to be supported by a particular UI element. The task with the highest priority should guide the design decision. However, a compromise may need to be found between design optimisation of a UI element supporting different tasks. On the other hand, a particular user task could also be supported by several alternative UI elements. For example, to support the user to access content items a video multi-screen can be offered together with a menu and an index, thus offering alternative accesses to choose from.

Users do not think in tasks but rather in goals they wish to accomplish. User tasks or task sequences serve user goals without being consciously selected to be carried out. Nevertheless, the identified generic user tasks for iTV applications can be combined in different ways to make up diverse user goals. Thus the presented interaction design solutions to support these user tasks can be used to support a variety of user goals.

Future Trends

Further research is needed to strengthen the empirical evidence that the identified design solutions are in fact proven solutions. Also, proven design solutions need to be identified for more user tasks by usability testing to offer more comprehensive design guidance on iTV-specific usability problems. It has been assumed that user task-based interaction design guidance supports a user-centred application development process. However, the final value of the design guidance would need to be evaluated in regard to the usability of the applications developed using the guidance compared to developing the same applications without the guidance. Also, the ease of use of the design guidance itself would need to be evaluated by iTV designers in regards to its use in the application development process. Our next step is the documentation of the identified design solutions in the form of interaction design patterns to better serve the user-centred development process of iTV applications.

Furthermore, technological developments offer new possibilities for user interaction that need new interaction design guidance. Advancements of the application standards promise many challenging interaction design problems. For example, Motion Picture Experts Group (MPEG)-4 based applications allow for user interaction with individual video and audio objects. Also, new interaction technologies, like pointing devices and other innovative remote control types will be accompanied by the need for new interaction design guidance.

Conclusion

The user task-based approach to identify proven interaction design solutions for iTV applications has proven useful. The identified generic user tasks formed a suitable basis for the usability test concept. The applied usability test methods are suitable to identify proven interaction design solutions. The results of this study provide some first design guidance. The final value of the presented results, however, depends on the experience and applied common sense of the iTV designer to make the best of it.

References

Brecht, R., Kunert, T., & Krömker, H. (2005). ISO standards zu Software-Usability und ihre Anwendbarkeit auf interaktive digitale Fernsehapplikationen (in German). In P. Scharff (Ed.), *Proceedings of the 50th International Scientific Colloquium of the Technical University of Ilmenau* (pp. 495-496). Ilmenau, Germany: Technische Universität Ilmenau.

British Broadcasting Corporation (BBC). (2006). *Designing for interactive television v1.0, BBCi & interactive TV programmes*. Retrieved December 11, 2006, from http://www.bbc.uk/guidelines/newmedia/desed/itv/itv_v1_2006.pdf

Dumas, J. S. (2003). User-based evaluation. In J. Jacko & A. Sears (Eds.), *The human-computer interaction handbook: fundamentals, evolving technologies and emerging applications* (pp. 1093-1117). Mahwah, NJ: Lawrence Erlbaum Associates.

Gawlinski, M. (2003). *Interactive television production*. Oxford, UK: Focal Press.

International Organsiation for Standardization (ISO). (1996-1998). *Ergonomic requirements for office work with visual display terminals (VDTs)* (ISO 9241).

International Organsiation for Standardization (ISO). (1996). *Ergonomic requirements for office work with visual display terminals (VDTs)—Part 10: Dialogue principles* (ISO 9241, Part 10).

International Organsiation for Standardization (ISO). (1998). *Ergonomic requirements for office work with visual display terminals (VDTs)—Part 11: Guidance on usability* (ISO 9241, Part 11).

International Organsiation for Standardization (ISO). (2002-2003). *Software ergonomics for multimedia user interfaces* (ISO 14915, Parts 1-3).

International Organsiation for Standardization (ISO). (2006). *Software engineering - software product requirements and evaluation (SQuaRE)- common industry format for usability test reports.* (ISO 25062).

Krömker, H., & Klimsa, P. (2005). Fernseh-Produktion (in German). In H. Krömker & P. Klimsa (Eds.), *Handbuch Medienproduktion* (pp.102-107). Wiesbaden: VS Verlag für Sozialwissenschaften.

Kunert, T., & Brecht, R. (in press). User requirements and design guidance for interactive TV news applications. In T. A. Rasmussen & J. F. Jensen (Eds.), *Interactive television: The media landscape, the users & the applications.*

Lu, K. Y. (2005). *Interaction design principles for interactive television.* Unpublished masters thesis, Georgia Institute of Technology, Atlanta.

Mayhew, D. (1999). *The usability engineering lifecycle.* San Francisco, San Diego, New York: Morgan Kaufmann.

Molich, R., & Nielsen, J. (1990). Improving a human-computer dialogue. *Communications of the ACM, 33*(3), 338-348.

Nielsen, J. (1993). *Usability engineering.* Boston, San Diego, New York: Morgan Kaufmann.

Pemberton, L., & Griffiths, R. N. (2003). Usability evaluation techniques for interactive television. In J. Jacko & C. Stephanidis (Eds.), *Human-computer interaction: Theory and practice. Proceedings of HCI International 2003.* Mahwah, NJ: Lawrence Erlbaum.

Preece, J., Rogers, Y., Sharp, H., Benyon, D., Holland, S., & Carey, T. (1994). *Human-computer interaction.* Wokingham, England, Reading, MA: Addison-Wesley.

Rinnetmäki (2004). *A guide for digital TV service producers.* ArviD publication 02/2004. Ministry of Transport and Communications, Finland. Retrieved November 29, 2005, from http://www.arvid.tv

Redish, J., & Wixon, D. (2003). Task analysis. In J. Jacko & A. Sears (Eds.), *The human-computer interaction handbook: Fundamentals, evolving technologies, and emerging applications* (pp. 922-940). Mahwah, NJ: Lawrence Erlbaum Associates.

Shneiderman, B., & Plaisant, C. (2005). *Designing the user interface—Strategies for effective human-computer-interaction* (4th ed.). Boston: Pearson Addison-Wesley.

Chapter XII

Guidelines for Designing Easy-to-Use Interactive Television Services:
Experiences from the ArviD Programme

Ari Ahonen, Technical Research Centre of Finland, Finland

Laura Turkki, Adage Usability, Finland

Marjukka Saarijärvi, Mininstry of the Interior, Finland

Maria Lahti, Technical Research Centre of Finland, Finland

Tytti Virtanen, VTT Technical Research Centre of Finland, Finland

Abstract

Ease of use is a key factor driving the adoption of interactive services in digital television (DTV). To facilitate the development of easy-to-use services and to distribute information among the industry players and the academia, the Finnish Ministry of Transport and Communications and the National Technology Agency of Finland commissioned a usability study on 11 DTV services. This paper presents the most topical results of these usability studies. First, we discuss the practical

experiences of evaluating differing interactive television (iTV) services. Second,
we present 39 guidelines for ensuring the ease of use of interactive services. These
guidelines apply to services that are transmitted in the traditional broadcast sys-
tem, but they also provide a good basis for designing services in Internet protocol
television (IPTV).

Introduction

The television is an everyday item for most people to such a degree that it is hardly
experienced as technology. The interactive services that DTV now offers, however,
constitute a relatively novel medium for consumers in which to use services and
find information.

A major factor that influences audience response to interactive services is ease of
use. The effort of grabbing the remote control and experimenting with new services
is lessened a great deal if the user is welcomed by an intuitive and friendly interface
designed from the user's point of view.

Ease of use has long been recognised as one of the major factors influencing the
adoption of new technologies, for example, in the technology acceptance model
(TAM) framework (see e.g., Davis, 1989; Kaasinen, 2005) and in the innovation
diffusion framework, in which it is approached from the perspective of complexity
(Rogers, 1995).

ArviD: A Cluster Programme Promoting DTV

In March 2004 the Finnish Government made a policy decision according to which
all television broadcasts should become digital in Finland from August 31, 2007
onwards. In order to facilitate a smooth transition to DTV and to promote the adoption of related interactive services, the Ministry of Transport and Communications
launched ArviD, a two-year cluster programme (2004-2005). The programme
sponsored the development of service innovations, ranging from entertainment to
business applications, as well as created a network for cluster-wide collaboration
and worked to achieve synergies in service production.

One of the main objectives of the programme was to promote the development of
interactive services for DTV. Fourteen service development projects were carried
out in 2005, based on an open call in the ArviD cluster programme (autumn 2004).
Broadcasters; network operators; public and private service providers; consumer
interest groups; and technology developers took part in these projects. The technical implementation of the services was based on the multimedia home platform

(MHP), which has been adopted as the common open standard for implementing iTV services in Finland.

Ease of use was recognised as a key factor for service adoption, and so it was agreed that usability evaluations of these service development projects would provide knowledge for design guidelines that would benefit the whole cluster. As a result, usability evaluations were conducted for 11 ArviD projects in 2005.

The usability studies produced hands-on experiences of evaluating iTV services, and most importantly, findings on how to design easy-to-use services. These findings are reported in respective sections in this chapter. The results have been reported more extensively in Finnish by Turkki et al. (2005).

Table 1. The evaluated services

Service	Evaluation method
Communal learning experience through DTV. *Developers: YLE (Finnish Broadcasting Company), Axel Technologies*	Expert evaluation and observation in simulation environment
Sign language in t-learning. *Developers: Helsinki University of Technology, Finnish Association for the Deaf, Ortikon Interactive*	Expert evaluation in an IPTV environment
Capital of Finland on DTV. *Developers: City of Helsinki, Sofia Digital*	Expert evaluation in broadcast environment
Interactive services of Akuutti health programme. *Developers: YLE (Finnish Broadcasting Company), Sofia Digital*	Expert evaluation in broadcast environment and user interviews
iTV Capture—Personalised screen capture. *Developers: MTV3, Infocast*	User tests in broadcast test environment
DTV services for the elderly and the disabled. *Developers: City of Oulu, Ortikon Interactive*	Expert evaluation of screen layouts
Library services on DTV. *Developers: Kirjastot.fi/Libraries.fi, Sofia Digital*	User tests in broadcast environment
EventTV—Box office in the living room. *Developers: Mix Media, Axel Technologies, Valimo Wireless*	User tests in broadcast environment
LumiTools—Multichannel publication service. *Developers: Lumi Interactive, Suomen 3KTV*	User tests in emulator environment using the wizard-of-oz method
SignText Context Subtitling. *Developers: Prosign, Finnish Association for the Deaf, Icareus*	Expert evaluation in broadcast test environment
Real-time public transport information on TV. *Developers: MTV3, Helsinki Public Transportation Authority, WSP LT Consultants*	User tests in broadcast test environment

Research Methods

The Evaluated Services

Altogether 11 services produced by the ArviD development projects were evaluated. Depending on the nature of the service and on technical factors, different evaluation methods were used for individual services (see Table 1).

Evaluation Methods

Traditional usability tests were conducted on five services. The tests were carried out in well-equipped usability laboratories. Two or three usability experts conducted the evaluations, and the process followed the standard usability evaluation practice, outlined for example in Nielsen (1993, p. 165-205). In each usability test, four users used the service, performing the main tasks for which the service was designed.

Expert evaluation was made on six services. In each expert evaluation two or three experts evaluated the service. Two services were analysed as fully operational services in a commercial broadcast environment. Three services were evaluated in a simulated or IPTV environment, and one service was evaluated as a paper prototype.

To provide richer data, observation and interviews were used in two cases as an additional method. In the observation case, free interaction between the user and the service was observed by two usability experts. Altogether 24 consumers took part in the evaluations or were interviewed. Most users participated in two service evaluations.

Analysis of the Results

Findings from individual evaluations related to ease of use were brought together for further analysis. The same experts that conducted the usability tests performed the analysis. The findings were grouped into emergent categories that reflected a general area of service design or an aspect of the user interface.

A set of design guidelines was then developed from the findings to cover each of the categories. The findings and drafted guidelines were compared with existing guidelines on the design of DTV services (e.g., British Broadcasting Corporation Interactive [BBCi], 2002; Carmichael, 1999; Rinnetmäki, 2004; Royal National Institute for the Blind [RNIB], 2005) and in addition, with DTV technical documents (European Broadcasting Union [EBU], 2005; Holappa et al., 2005) as well as with well-established general usability guidelines and Web design guidelines (e.g., Ala-Harja & Lindh, 2004; Chisholm, Vanderheiden, & Jacobs, 1999; Koyani, Bailey, & Nall, 2003; Nielsen, 1993).

This resulted in an initial set of guidelines, which were then submitted for comments to the DTV service developer community within the ArviD programme. Based on the comments from service developers, a final set of 98 guidelines with 182 associated checkpoints was reached (Turkki et al., 2005). To create a more compact synthesis, these were further distilled into the 39 guidelines reported in this article.

Practical Experiences of Evaluation

The core of conducting usability evaluations on DTV services does not differ from usability studies in any other environment. However, the characteristics of the broadcast environment (see e.g., Morris & Smith-Chaigneau, 2005) and the rapidly evolving nature of the technical infrastructure create some challenges that need to be addressed or at least prepared for. Based on the experiences of evaluating ArviD services, the following aspects are suggested as important practical factors that should be considered.

First, only a real broadcast environment lets users and experts assess the delays of the service realistically. Improved developer versions of the set-top boxes that allow for rapid testing are useful, but they are also unrealistically fast, because the delays of the broadcast system, most notably of the object carousel, are not present. Since slowness is one of the most common complaints about iTV services, it affects the user experience even when testing.

Second, the service may be available for testing in a real production environment only for short periods of time when it is being broadcast. Programme-related interactive services may be available only once a week when the weekly episode is on the air.

Third, it is difficult to get the unfinished prototypes into laboratory. The usability evaluations may need to be conducted at the developers' premises. Portable usability laboratories, if available, are useful in this regard.

Fourth, the set-top boxes can be different not only in terms of performance but also in terms of implementation of the middleware. Several set-top boxes should be used for testing in order to observe the functioning of the service in different end-user devices.

Fifth, computer-based environments (e.g., MHP emulators or browser environments) can be used for early usability testing. These emulators can present both functionality and structure correctly, but interacting with the computer is fundamentally different from interacting with the television. The distance to the screen is different; the colours and the resolution are not experienced correctly; and the interaction is not the same as the real interaction using a remote control. For evaluating services in emulator environment, applying the "wizard-of-oz" method (Gould, Conti, & Hovanyecz, 1983; Kelly, 1984) can prove to be valuable. It can be utilised to make the user experience more realistic and fluent: The screen contents of the emulator can be directed to the television screen, and the evaluators can then ask the user which remote control keys he or she would press. The wizard, who is actually controlling the service with the computer emulator, can then apply the commands as if the service were responding to the remote control keys indicated by the user.

Sixth, before the digital services become everyday commodities, it may be difficult to find test users from the target group who are familiar with iTV services. Therefore, it is useful to let the users learn the basic operating principles with other DTV services before the actual test is started. For example, digital teletext can serve as a good learning ground. Also, in some cases testing more than one service with the same test users can be beneficial. This can be done if the target groups of the services are similar and the services can be tested outside the development environment. This will provide users with even more experience, and they can concentrate on the service in question rather than the basic operating principles. Should more than one service be tested at the same time, it is necessary to control the order of the tested services across the users.

The Guidelines

The intended audience for the following guidelines are the interactive service developers. The guidelines focus on the ease of use, hence they do not provide detailed instructions on visual design or technical implementation.

The guidelines have been structured from a pragmatic viewpoint into five design areas: (1) concept definition, (2) functional design, (3) structural design, (4) look and feel design, and (5) instructional design. Concept definition takes place in the very beginning of the design process but the work in the other design areas usually proceeds more or less in parallel.

The guidelines are meant to help a proactive design for ease of use. However, it goes without saying that these guidelines cannot substitute iterative usability evaluations. From early on, sketches, visualisations, and models that come out of the design process should be evaluated with users, usability professionals, or if no one else is available, with one's colleagues who are not involved with the design.

In addition to the ArviD usability evaluations, the guidelines presented here build on earlier recommendations and guidelines. To keep the text readable, the detailed sources to each guideline are not indicated in the following passages. Especially important literary sources have been Rinnetmäki (2004) and BBCi (2002).

Concept Definition

Concept definition creates the foundation for the whole design by setting the requirements and defining the users' goals that the service must serve. Especially in the concept definition phase cooperation is needed: Service developers should discuss and resolve certain design aspects together with broadcasters and programme producers.

Know the user, the context of use, and the user's goals. When defining the concept, make sure that you know who the user is and that you understand the user's goals and the context of use, that is, how and where the service is used and why. Establish information as to what the users already know and what they need to learn before using the service in its context. At an early phase, consider the need for accessible design.

Make sure that the service has an edge that catches the user's interest. The service needs to have a raison d'être, a competitive advantage over similar services in other media. DTV must be an especially suitable platform for the service, offering users something that other media cannot offer.

Pursue unity within larger context. Design the service in line with the broader service concept, if such exists. For instance, functional implementation and the look and feel of the service should comply with a programme format, channel brand, or cross-media service concept if such are associated with the service.

Use conventions when available. Make sure that the service is in line with any conventions that have been adopted in the target market, for example, displaying colour key functions always at the bottom of the screen, or using the same colour keys for certain functions. If you decide to break the convention, make sure to test the feasibility of the solution with users. Different markets have different conventions, so identify which markets are important from your perspective.

Make sure that the users know how to begin to use the service. Make sure that the service is readily available and that users can find it. Together with the broadcaster, make sure that users understand the on-screen impulse icon signalling service availability. Make sure it is clear to users whether the service is connected to a broadcast programme.

Make sure that there is enough space for interactive elements on the screen. If the service is associated with a television programme, make sure that there is enough room for interactive elements in the visual stream. The use of screen space needs to be planned simultaneously with planning the shooting of the programme.

The programme must be viewable with and without interactive elements on the screen.

Functional Design

Designing functionality, that is, what the user can do with the service and how, starts where concept definition ends. Functional design determines how the service works and how it responds to user commands.

Minimise delays. The pleasantness of use depends greatly on the speed of the service and on how quickly the user gets feedback for his or her actions. The basic navigation should respond to the user in less than 1 second (BBCi, 2002). If the delays are longer, the user should be given a clear indication of the delay. If possible, try to load information into the set-top box while the user is busy doing something else, for example, registering to the service.

Note the possible limitations of the browser environment. If you design for a browser environment, note that the browser can impose limitations on service implementation. If you want transferability to more than one browser, do not place important functionality behind the colour keys; chances are that other browsers cannot show them.

Ensure easy crossover to other media. Some services require the use of additional media, for example, mobile phones, to complete service transactions. Make sure that the crossover from the user interface in television into using another medium is intuitive and well instructed. The user should receive sufficient feedback via the most natural channel, which usually is the medium used in the interaction. The critical moment in the crossover is when the user should understand to change the medium and most of all, have the motivation to do so.

Use only those keys as controls that exist on every remote control unit. Design the service so that it is operated only with the keys that can be found in each remote control type (as defined, for example, in NorDig (2005, pp. 75-76). Use non-standard, extra keys only to provide additional functionality like shortcuts. Indicate the keys in use clearly in the user interface on screen.

Give the users enough time. First, provide the user with options to control any multimedia material that the service provides. Let the user decide when to watch videos or listen to audio files and provide a possibility for pausing. Do not use pre-determined timing for presenting the material unless you have a compelling reason for it. Second, make sure that users have enough time to interact with the service: For example, it takes more time for a group of people to decide a joint answer to an interactive quiz show than for an individual.

Display the status of the interactive service clearly. If the service has different modes, for example, for inputting text with a virtual keyboard, for browsing instructions, or for performing actions indicate the mode and its changes clearly. Also make sure that the user knows the status of the return channel and understands if there are delays in the use of the service.

Make sure that the user is aware of all limitations and requirements related to the use of the service. Make sure that the user understands if there are costs in using the service, whether the service requires a return channel, and whether the service can feasibly be used with the type of set-top box or DTV set that the user has.

Make sure the content is valid. Make sure that the content is valid and up-to-date at all times. Make sure that any automatically generated content is understandable, that headings are comprehensible, and that layout is not broken. Remember that while automatic processes may enable easy content production, it does not equal with easy content comprehension.

Make text input easy. The need for inputting text should be minimised in the first place unless you can be sure that all users have access to writing devices such as infrared keyboards. If possible, let the user select alternatives from a list rather than input information as text. If the user needs to input text, make sure that he or she knows how to start typing and that there is enough time for it. Make sure that the user knows which information is compulsory and that the user does not loose information even if he or she needs to go back in the data gathering process. To prevent errors, give instructions on the expected input format, for example, ddmmyy. If a virtual keyboard is used, make sure it is well integrated to the service. The virtual keyboard should always open in the right mode and follow the language and the look and feel properties of the service. Additionally, make sure it is easy for users to remove or correct input errors, with or without a virtual keyboard.

Let the programme sound continue in the background unless there is a good reason to silence it. The programme sound lets the user follow what is happening in the programme stream even if the visual information space is taken up by the interactive service.

Provide users with accessibility features. Provide users with accessible settings, for example, possibility to change text size or use plain language. Make sure that accessibility features are easily found.

Structural Design

Structural design means dividing the information and functionality into a suitable hierarchy and understandable blocks that can be displayed on the screen and creating a way to navigate within the service. The design of the structure encompasses the decisions about what content and functions to place on individual pages (i.e., how to build the page hierarchy and define interconnections between the pages) and also how to structure the contents on a given page (page layout).

Place the most important functions up front. The main task is to identify the top-priority functions and information and to ensure that the user has an easy access to them within the fairly limited DTV interface. Offer access to the most important functionality on the initial page and place important functions up front on every page.

Make the structure as simple as possible and display it clearly. A simple structure helps the users comprehend the idea of the service and feel in control of it. Indicate the user's location in the service on screen. Also indicate how the user can navigate within a page or move from one part of the service to another.

Make sure it is easy to exit from the service. Make sure that the user can easily return to watch the broadcast.

Make sure that the cursor moves logically. Do not let the cursor jump over links. On each page make sure that the most important links are close to the initial position of the cursor. With visual cues like proximity, indicate to the user which links belong together. Clearly highlight the active element on which the cursor (or focus) is.

Give textual content a structure that is easy to read and easy to comprehend. Divide text into easy-to-read paragraphs and use headings for making the text easy to browse. Make sure that information is not separated from headings even when the content is automatically produced. Use lists to make textual content easy to browse and quick to read. Arrange large textual content into deep hierarchical structures.

Make navigation systematic. Make sure that the navigation throughout the service is systematic. There are six common basic ways of navigating within the service (Turkki et al., 2005):

- Moving the cursor with arrow keys and selecting with the OK key
- Using left and right keys to change the screen content without the OK key
- Using the arrow and the OK keys to select both functionality and links on the same page
- Using the arrow key to move the content while the cursor stays in a fixed place
- Using colour keys for navigation
- Using number keys as link shortcuts

The methods are suited for different situations. Try not to mix different navigation methods, at least not on the same page. Note: remember that the users cannot see the options that are placed in a menu behind a colour key. Many times it is simpler to use colour keys for simple functionality, for example, options such as "previous page" and "initial page."

Make the layout and choice of remote control keys consistent. If links or options are placed on the screen vertically, use the up and down arrow keys to scroll through them. If the options are placed horisontally, use the left and right arrow keys. Navigation elements should be easily distinguishable from other elements.

Look and Feel Design

The look and feel of the service brings the structure, functionality, and the style of the service together both visually and experientially.

Display the active status of the service clearly. Ensure that the user understands when the service is active and when remote control functions are used to control the service instead of changing channels, for example.

Indicate visually which elements are similar in function on the television screen. Make sure that all user interface elements of the same type are marked in a consistent way throughout the service, for example, all link elements share visual features that identify them as links. At the same time, make sure that different kinds of elements are distinguishable from each other, for example, links differ from plain text. Note: since most services are navigated with arrow keys, arrow-shaped elements should be used on screen only to indicate the use of arrow keys of the remote control (and not, for example, as list markers).

Show all on-screen graphical elements at the same time. If visual elements appear on screen one by one, it creates visual clutter, and the user may not know when the page is complete. Eliminating the unnecessary distraction created by a piecemeal loading of graphics is especially important if there is material that starts automatically on screen, for example, video clips. However, make sure that the user does not have to wait too long looking at an empty screen.

Ensure the readability of the text. Use a suitable sans serif font, like Tiresias screen font, and use only recommended font sizes. BBCi (2002) recommends 24 points as a minimum for normal text and 18 points as the absolute minimum size, RNIB (2005) recommends 24 points as a minimum, and the MHP specification (EBU, 2005, p. 532) provides use cases for font sizes 24-36. Take the target audience into account when selecting font sizes. Also make sure that there is adequate contrast between text and the background and see to it that background images do not make it difficult or impossible to read the text.

Make sure that all information conveyed with colour is also understood without colour. All functional information must be understandable without the use of colour, especially reds and greens with similar colour saturation.

Design for television screens and formats. Take into account differences in aspect ratios, the traditional 4:3 and wide-screen 16:9, and differences between standard and high-definition television. Also, leave safe margins on the screen, because all television sets do not necessarily show all the pixels on the edges. Make sure your graphics look good on television resolutions and the available colour palette(s).

Make sure that the relationship between the service and the broadcast is clear. Make sure that users know which elements belong to the service and which are part of the broadcast on the background.

Be sparing in showing the television programme stream in the service. The moving content of the television video stream may hamper users because their attention is easily distracted by the movements. Another solution is to let the user opt to show or hide the programme. This is especially important when the service is not connected to the programme shown in the background.

Use language that is familiar to the target audience. Use language that is suitable for the users. Avoid technical terms and explain them if used. Make sure that the labels of links and buttons make sense to the users and present the actions clearly.

Show the colour key symbols on the screen constantly. The instructions for colour function keys should be constantly displayed on the screen, in the same order as the buttons appear in the remote control, even when their functions are not in use. This makes it easier for the user to recognise and understand the meaning of each colour key. If some key is not used on a given page, indicate this for example by dimming the key description.

Instructional Design

Ease of use is largely based on the ability of the user interface to convey information on how the service can be used. Nevertheless, in the adoption phase of iTV and novel services, there is a pronounced need for comprehensive user guidance.

An interactive service usually contains two kinds of on-screen instructions: (1) embedded operating instructions in the user interface, and (2) a separate set of instructions in a guide section, which may provide more information on how to use the service.

Make sure that instructions are understandable by testing them with actual users. Terms that belong to engineers' everyday language are often unknown to the users. Also, instructions are usually written by persons who are familiar with the service. This familiarity may lead them to omit important information that they have taken for granted. For this reason, always test with the users that the instructions are understandable.

Make the purpose of the service clear on the initial page. Provide concise information on the opening screen about what the user can do with the service. The user may not have a clear prior understanding of the service, or may even have opened the service by an accidental key press.

Display the most important instructions in the user interface. Intended or not by designers, all information on screen and even empty spaces around objects serve as a clue to how to use the service. Use informative headings and breadcrumb trails to tell the user his or her location in the service. Do not simply trust that the user will guess or try out which key to press—indicate clearly which keys should be used. Provide enough instructions on how to fill in fields or how to use other interactive elements. Also, tell the user how much the use of the service costs.

Make separate user instructions easy to read and quick to browse. On the television screen, large amounts of text are hard to read. Therefore, arrange the user manual in a deep hierarchy where the user can easily browse through the content. Usually a good method is to arrange the user manual to respond to the users' goals, for example, by formatting the manual headings into a "how to" format, for example, "how to open the return channel," "how to order tickets."

Tell the user who to contact if more help is needed. Make sure that the user knows where to get extra help on the Web or by phone. This also works as a feedback loop to the service developer.

Discussion

On Guidelines

The guidelines presented in this paper are based on the results of usability evaluations as well as on previous guidelines. As often is the case with usability evaluations, the sample sizes in the user tests were small and the obtained material mostly qualitative in nature. As for the expert evaluations, they are by definition based on the expertise of the evaluators. While this kind of data enables a detailed understanding of the issues involved, it is hard or impossible to provide hard scientific evidence for most of the guidelines.

Therefore, it is best to approach the guidelines from the perspective of practical value: If they are beneficial for the creation of easy-to-use services, they are valid. Many guidelines touch on topics that would warrant meticulous scientific research. Such topics are, for example, the effects of delays on user satisfaction and the minimum font sizes for different audiences. The research would greatly benefit both practitioners and academics involved with iTV services. The greatest beneficiaries would be the end users.

While guidelines can be of great benefit, they never substitute open discourse with the members of the targeted user group. Designing good interactive services requires both proactive planning and the continuous collection of critical feedback from users, usability experts, and subject matter experts. Another thing that the guidelines cannot replace is the will to try to solve the problems.

The guidelines presented here are concerned with ease of use, but ease of use is not the whole picture; it is merely a precondition for successful services. As Eronen (2004, p. 86) points out, with a television interface one needs to go beyond the ease of use or usability and take into account aesthetic qualities. It is difficult to give guidelines on how to design for an aesthetically pleasing experience, but every effort should be made to ensure that the interface pleases the eye of the beholder.

On Challenges

Design guidelines are always written with a presupposition that there are potential challenges lurking ahead. What is it, then, that makes designing easy-to-use iTV services challenging in the first place? Different service types have their own challenges and difficulties, but they also share a large amount of common ground because of the underlying television platform. The common problems stem from two sources: (1) the interaction paradigm based on a low-resolution screen and a remote control, and (2) the immaturity of the technological solutions.

Some usability issues are inherent in the interaction model based on television and remote control. Complex, computer-style interaction is simply not possible (that is, without a mouse). Therefore, it would seem logical not to try to do too much with this limited interface: A complex iTV service would appear like an unintended joke in comparison with the same service over broadband Internet, used smoothly with the mouse and cursor running over those countless pixels on the computer monitor. Even if the user does not have previous experience in the computer world, his or her intention to use a service will diminish under complexity. At the moment, fairly simple services have the best chances of being adopted.

Certain other challenges are directly related to the underlying technologies, which are to a large degree immature. The return channel implemented with a conventional telephone modem, commonly used in set-top boxes today, was out of date from the very beginning. The excruciating wait for the modem to connect must be experienced in person in order to understand how long it takes. Set-top boxes have very limited computational power, which makes complex graphical or functional operations difficult. The broadcast of applications and service information via the object carousel, together with a limited amount of memory in the set-top box, results in long load times and delays. This is the current situation, but the immaturity also holds a promise: at least things can improve.

The Road Ahead

There are new challenges on the horison for designing easy-to-use services. The development of the DTV platform will lead into a more stable and effective operating environment, thus improving performance. However, new developments will also enable new service concepts that bring along a new set of questions to be addressed.

In the traditional broadcast television model, a broadband return channel will change many things in offering interactive services. Things will run more smoothly, and finally e-government and transaction services may become reality. This will give privacy and security design issues a more central role than what they hold at the moment. Another, somewhat smaller change will come if wireless keyboards become more common at the side of the remote control unit. They will make text-based interaction easier, and new service concepts will become feasible; for example, chatting with instant messages among a group of friends while watching the same programme in different locations. Higher-resolution displays with larger screen sizes will also eventually bring more degrees of freedom to the design of graphics and layout, even though legacy devices will remain at users' homes for years to come.

At the same time, the traditional broadcast model is challenged by two potentially revolutionary services or products that are emerging on the market: IPTV and home media centre solutions. These two combined with traditional services provide a plethora of new service concepts. Combining the respective strengths of the broadcast and unicast (point-to-point communication) will be one of the most topical questions. Because of this development, in a few years it will be necessary to write a new set of guidelines, not so much from the perspective of technical limitations, but from the perspective of how to provide complex services in intuitive packages.

As to the guidelines presented here, even if the wording of some guidelines will soon make them obsolete, the basic principles should hold. This is especially true on the basic message underlying the guidelines. The message is "make it simple." Humans as users or as television watchers are able to do incredibly complex things but they do not necessarily like it. People like to use simple things.

Acknowledgment

The authors wish to thank the persons working with the ArviD services for their kind cooperation and for contributing many insights into the nature of iTV. In particular, our thanks go to Mikael Rinnetmäki from Sofia Digital and Arttu Heinonen and Rami Aalto from Ortikon Interactive for their time and valuable comments. Pekka Nykänen and Arto Saikanmäki from JP-Epstar had a fundamental in role in the process of producing the guidelines. The research was funded by the National Technology Agency of Finland, Tekes, together with the Ministry of Transport and Communications, which is gratefully acknowledged.

References

Ala-Harja, M., & Lindh, C. (2004, August). *Quality criteria of public web services* (in Finnish). Working party memoranda. Helsinki, Finland: Ministry of Finance.

British Broadcasting Corporation Interactive (BBCi). (2002). *Interactive television style guide*. BBCi White Paper. Retrieved August 11, 2005, from http://www.bbc.co.uk/commissioning/newmedia/itv.shtml

Carmichael, A. (1999). *Style guide for the design of interactive television services for elderly viewers*. Retrieved November 21, 2005, from http://www.computing.dundee.ac.uk/projects/UTOPIA/Publications.asp

Chisholm, W., Vanderheiden, G., & Jacobs, I. (Eds.). (1999, May 5). *Web content accessibility guidelines 1.0*. Web Accessibility Initiative. Retrieved November 21, 2005, from http://www.w3.org/TR/WCAG10/

Davis, F. (1989). Perceived usefulness, perceived ease of use, and user acceptance of information technology. *MIS Quarterly, 13,* 318-340.

Eronen, L. (2004). *User centered design of new and novel products: Case digital television*. Publications in telecommunications software and multimedia. Espoo, Finland: Helsinki University of Technology.

European Broadcasting Union (EBU). (2005). *Digital video broadcasting (DVB); Multimedia Home Platform (MHP) Specification 1.1.2*. The DVB Project. Retrieved November 28, 2005, from http://www.mhp.org/mhp_technology/mhp_1_1/

Gould, J. D., Conti, J., & Hovanyecz, T. (1983). Composing letters with a simulated listening typewriter. *Communications of the ACM, 26*(4), 295-308.

Holappa, J., Ahonen, P., Eronen, J., Kajava, J., Kaksonen, T., Karjalainen, K., et al. (2005). *Information security threats and solutions in digital television*. VTT Research Notes 2306. Espoo, Finland: VTT Technical Research Centre of Finland.

Kaasinen, E. (2005). *User acceptance of mobile services—Vvalue, ease of use, trust, and ease of adoption*. Doctoral dissertation. VTT Publications 566. Espoo, Finland: VTT Technical Research Centre of Finland.

Kelley, J. F. (1984). An iterative design methodology for user-friendly natural language office information applications. *ACM Transactions on Information Systems (TOIS)*, *2*(1), 26-41.

Koyani, S., Bailey, R., & Nall, J. (2003). *Research based web-design & usability guidelines*. Retrieved August 11, 2005, from http://usability.gov/guidelines/guidelines_notice.html

Morris, S., & Smith-Chaigneau, A. (2005). *Interactive TV standards*. Burlington: Elsevier/Focal Press.

Nielsen, J. (1993). *Usability engineering*. San Diego, CA: Academic Press.

NorDig. (2005). *NorDig Unified Requirements for Integrated Receiver Decoders*. Retrieved March 15, 2006, from http://www.nordig.org/pdf/NorDig-Unified ,%20ver%201.0.2.pdf

Rinnetmäki, M. (2004, February). *A guide for digital TV service producers*. ArviD publications. Helsinki, Finland: Ministry of Transport and Communications.

Rogers, E. (1995). *Diffusion of innovations* (4th ed.). New York: The Free Press.

Royal National Institute for the Blind (RNIB). (2005). *Guidelines for the design of accessible information and communication technology systems: Television*. Retrieved November 21, 2005, from http://www.tiresias.org/guidelines/television.htm

Turkki, L., Ahonen, A., Lahti, M., Virtanen, T., Saastamoinen, M., Vastamäki, R., et al. (2005, July). *Design guide for usable digital television services* (in Finnish). ArviD Publications. Helsinki, Finland: Ministry of Transport and Communications.

Chapter XIII

Text Editing in Digital Terrestrial Television:
A Comparison of Three Interfaces

Arianna Iatrino, CSP Innovazione nelle ICT S.C. a r.l., Italy

Sonia Modeo, CSP Innovazione nelle ICT S.C. a r.l., Italy

Abstract

This chapter introduces the usability problems regarding text entry using a remote control in digital terrestrial television (DTT) context. It describes the comparison of three different text editing interfaces: (1) the multi-press with timeout, (2) the multi-press with timeout and visual feedback, and (3) the virtual keyboard interface. This chapter describes a test based on a within-group design: The authors analyse the efficiency, the effectiveness, and the user's satisfaction of the three interfaces mentioned previously. The study shows a significant relationship between the users' level of experience in text editing using a mobile phone and their favourite interface. Moreover, the analysis demonstrates that there is no relationship between users' level of experience and the editing problems they encountered. The authors hope that this study will help interactive application developers in designing usable interfaces for text entry using a remote control.

Introduction

By the end of 2012, there will be the European migration from the traditional analogue broadcast system to DTT. This means that all television channels will be broadcasted digitally. Every country fixed a different date for the so-called "switch off." In Italy, for example, it will be at the end of 2008.

Besides the technical details, the main innovation will be the enhanced interactivity: By simply connecting a set-top-box to a television, users will be able to enjoy additional content and interactive applications. Furthermore, via a return channel, users will be able to send data to a broadcaster (for example, to participate in quizzes, to vote, to play games, and so on). For this reason public funding was released by the Italian government to support the new market in order to increase sales of interactive DTT set-top-boxes and to spread t-government services to all kinds of users: young and old, expert, and naive.

We can imagine a future scenario, in which, for example, an elderly person has to fill out records to book a medical examination by DTT. The only device she has to interact with the DTT is the remote control.

We can also suppose that young people are facilitated in editing text using a remote control because of their familiarity in sending messages via a short message system (SMS) with a mobile phone. Since DTT will involve an enlarged target, it is important to study criticalities in text editing both for expert and non-expert users. As a consequence it is fundamental to design a usable graphical interface for text editing.

The aim of this chapter is to evaluate three different text input interfaces in order to find out if there is an easy and fast way to edit text using a remote control.

This chapter is organised as follows: in the next section we present some related works concerning text input methods; then we describe the three different input interfaces that we evaluated and we detail the experimental design; and finally, we analyse the results and we present future trends and conclusions.

Background

The advent of interactive television (iTV) applications allows users to interact with TV and to send data to broadcasters. Sometimes, in order to enjoy additional contents (for example to read and send e-mails, to store games' score, or to buy something) users have to enter some kind of text (username and password rather than name, address, and so on). Currently, the remote control is the only input device available to enter text on DTT applications.

Even if text entry is a well know field in human-computer interaction (HCI) studies, text entry using a remote control is a novel issue. Several studies concerning methods for text entry have been undertaken within the HCI field (MacKenzie, 2002). However, most of the researched methods are not suited for the home entertainment context because they require input devices very different from a handheld DTT remote control, such as a stylus (Zhai, Smith, & Hunter, 2003) or a QWERTY keyboard (Karat, Halverson, Horn, & Karat, 1999). In particular, studies conducted on text entry on personal digital assistants (PDAs), computers, and mobile phones could not be applied in DTT context because of three reasons:

- **Input devices.** Computers, for example, have QWERTY keyboards while DTT has just a multi-press alpha-numeric keyboard, similar to mobile phones' one but with some differences.
- **Context of use.** In DTT context, the distance between user and device is higher than in other contexts such as on PDAs, computers, and mobile phones. Moreover, the distance between screen and input device is also higher in DTT. For that reason, the user's attention is divided between the screen and the remote control.
- **Set-top-box delay.** DTT set-top-box has a retard in showing output on the TV screen so users have some problems in understanding if they selected any character or if they did the correct number of presses the on remote control.

Despite the peculiarity in text entry using a remote control, the alpha-numerical keys of DTT remote control (see Figure 1) can be compared to a mobile phone keypad (see Figure 2) because they use similar input techniques.

Several works in text editing using mobile phones are significantly related to this project. For example, Pavlovych and Stuerzlinger (2004) studied a new model for predicting text entry speed on a 12-button mobile phone keypad. Silfverberg, MacKenzie, and Korhonen (2000) and Butts and Cockburn (2002) evaluated three text input techniques on mobile phones: (1) two-keys, (2) multi-press with timeout, and (3) multi-press with next button input methods. Subjective ratings of the three methods did not yield significant differences, and the experimental users rated all three text input methods frustrating.

An innovative study regarding text entry using a remote control was conducted by Ingmarsson, Dinka, and Zhai (2004). They tested the performance of a novel text input method for iTV: The numpad typer (TNT). Within TNT the TV screen shows letters and special characters chunked into six groups: The user has to press two numeric keys to produce a letter on the screen. The first one selects a group while the second one selects the character. Despite TNT, performance was comparable or superior to the current PDA handwriting or multi-tap methods, the study evaluated

Figure 1. Typical DTT remote control (DGTVi, 2004)

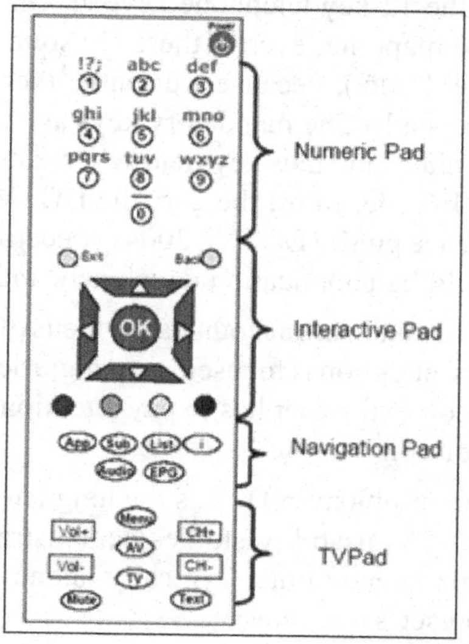

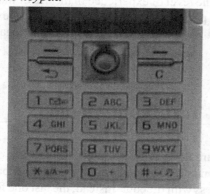

Figure 2. Mobile phone keypad

only a small group of users familiar both with QWERTY keyboards and T9 system on their own mobile phone.

Notice that the comparison between the remote control and the mobile phone keypad can be made but with some caution because of some differences.

First of all, one of the main issues when entering text on mobile phones is the mapping between letters and keys. Each key is mapped to at least three letters, following the European Telecommunications Standards Institute (ETSI) ES 202 130 V1.1.1 (2003) standard for character repertoires. This consists of a set of ordering rules and assignments to the 12-key telephone keypad. On a DTT remote control we found almost the same mapping, even if there are some differences regarding special characters (i.e., @, !, etc.), because, currently, there are no standards for remote control numeric keypads. The mandatory keys and key events available to the application are very limited, and thus keys and key events may vary from manufacturer to manufacturer. In order to fill the gap, the DGTVi society (http://www.dgtvi.it) published a reference guide (DGTVi, 2004) concerning DTT receivers and remote controls, aimed at Italian application developers and users.

Another issue that should be taken into account is the focus of users' attention. When using a mobile phone users' attention is focused on a single device (the mobile phone itself), while in the DTT context, a user has to pay attention both to the screen and to the handheld remote control.

Finally, a peculiar usability problem in DTT is the long latency time between user input and screen feedback. As usability studies demonstrate (Miller, 1968), it is fundamental to reduce this latency time: Typically latency should be less than 1 second in order to keep a user's attention.

Considering all the issues explained previously, we evaluated the usability of three interfaces in order to understand (1) if there is a more usable interface to enter text editing using DTT, (2) if the level of experience in the use of mobile phones to send SMS can influence users' performance, and (3) if the level of experience in the use of mobile phones to send SMS can influence preferences regarding the three text input interfaces.

Text Input Method

In this chapter we compare three different interfaces for text editing using a DTT remote control. The first and the second interface follow the mobile phone text entry paradigm called multi-press with timeout. Instead, the last one follows the keyboard typewriter paradigm. Each of these techniques is detailed next.

Multi-Press with Timeout Interface

In the multi-press with timeout technique, a user cycles through letters by pressing a single key several times. If the user does not press any key during a predefined time (timeout) the interface will select the character currently on screen. For example, to edit "CDE," the user must press the "3" key once, then wait for the timeout to expire. Then he/she must press the "3" key twice. Finally, once the timeout has expired, the user must press the "3" key three times.

Figure 3. Multi-press with timeout keyboard

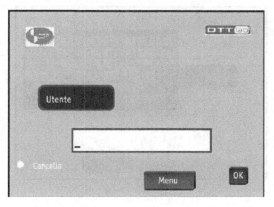

Figure 4. Multi-press with timeout and visual feedback keyboard

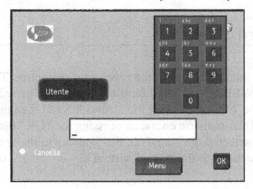

In our multi-press with timeout interface (see Figure 3), a text box shows the user the cycling of the alpha-numeric character mapped to the key that they are pressing. The cancel function is associated to the yellow button on the interactive keypad, represented by a yellow icon on the TV screen.

Multi-Press with Timeout and Visual Feedback Interface

This interface is a variation of the multi-press with timeout technique based on a representation of the numeric keypad of the remote control on the TV screen (see Figure 4). In this way the user gets a visual feedback regarding the character he/she

Figure 5. Virtual keyboard

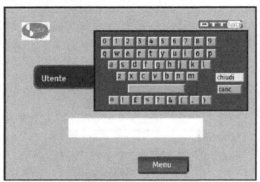

is selecting by watching the cycling of the alpha-numeric character both in the text box area and on the numeric pad representation.

Virtual Keyboard

The virtual keyboard interface follows the typewriter metaphor: There is a visual representation of a QWERTY keyboard on the TV screen. The user navigates the virtual keyboard on the TV screen using the arrows keys of the interactive keypad and selects the right alpha-numeric character pressing the "OK" button (see Figure 5). Once the user has selected the character, this appears in the text box area. The cancel function is associated to the "Canc" virtual key on the TV interface.

Experimental Design

The aim of this evaluation is to determine whether there are meaningful differences between the efficiency, the effectiveness, and the user's satisfaction of the three text editing interfaces. Regarding the efficiency, we calculated the average time spent to complete each task, the percentage of completed tasks, and the average number of characters typed per second (cps). Instead, the average number of mistakes made during the text editing was considered as a measure of effectiveness. Finally, questionnaires data were analysed to determine users' satisfaction.

The experiment was conducted using a within-group design (a design that repeats observations on the same subject), with the interface type as an independent variable.

The within-group design is set against the between-subject design. In a between-subject design, a subject is observed in one and only one treatment combination, while in within-group design it is possible to observe one subject in more than one treatment condition. We chose the within-group experimental design in order to have the same users' variance in the performances of the three interfaces and in order to decrease the effort of recruiting new users.

As mentioned before, the possible interface types were (1) the multi-press with timeout, (2) the multi-press with timeout and visual feedback, and (3) the virtual keyboard.

Thirty-six subjects (selected considering their level of experience in typing SMS) participated in the experiment solving six tasks (two with each interface). The order in which the users experienced the three interfaces as well as the order of tasks was random, to minimise learning effects.

The subjects were divided into two subgroups (18 users each), according to the different level of experience in entering text using a mobile phone. We considered as expert users those who send more than 5 SMS per week and as naive users all the others. The expert user group included 10 male and 8 female, aged 18-50 while the naive user group included 8 male and 10 female, aged 43-80.

Figure 6. Laboratory room

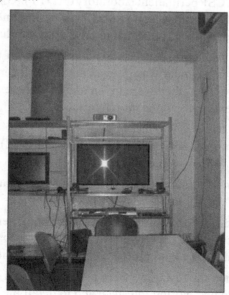

Figure 7. ADB Embox remote control

Procedure

The evaluation was conducted in the laboratory room (see Figure 6) of the DTTLab (http://www.dttlab.it is a CSP permanent laboratory on DTT)—one of the permanent research groups of the CSP research centre (http://www.csp.it/en/)—showing the interfaces on a TV (Medion—42") connected to a DTT set-top-box (ADB-Embox).

Users were given an introductory letter to explain the aim of the test and clarify that the purpose of the experiment was to evaluate three different text editing interfaces, not their ability. Then they were given a DTT remote control (see Figure 7) to interact with the text input application interfaces.

Each subject was asked to solve two tasks using each interface: (1) in Task 1 (T. 1) the user had to enter an e-mail address (pippo@libero.com); (2) in Task 2 (T. 2) the user had to type a short Italian sentence (*sole e neve*).

Naive users were given a training period of 5 minutes to familiarise themselves with the input method.

Each test session was video recorded in order to measure both user's performance and time spent to complete each task. Moreover, we asked users to express their thoughts and questions aloud (thinking-aloud protocol), to make a qualitative evaluation.

After the users completed the test, they were asked to fill in a short questionnaire to choose their favourite text editing interface and to detail the problems found. In particular, the questionnaire was divided into three sections. The first section concerned social-demographical data: age, gender, profession, and the average number of SMS sent per week. The second one had multiple-choice questions regarding the interaction: user's favourite interface, problems in the use of each interface, and so forth. The last section of the questionnaire was dedicated to users' comments and questions.

Results

The analysis showed that all the tested interfaces failed, mainly because of the delay of the set-top-box.

In particular, both the multi-press with timeout interfaces had these main problems:

- **Localisation of special characters key**. All non-expert users and two expert users needed help to find the symbols key. In our application all the special characters were associated to the "1" key like in many mobile phones. Even if users were told where the special characters were, most of them needed to be assisted in order to complete the task.

- **Number of key presses in order to select a special character**. Once they were told where a special character was, users speculated on how many times they had to press the key to find the required special characters ("@" and "."). In fact, special characters' order of appearance is not standard, neither on current mobile phones nor on DTT applications. All non-expert users stopped pressing the "1" key after three presses, while expert users cycled through the special characters until they found the right one.

- **Localisation of the cancel key**. Although a yellow icon on the TV screen advised the users that the cancel function was associated with the yellow button on the DTT remote control, most of the users (both expert and naive) needed help to find the cancel key. The problem was that the users just looked at the remote control even if, in order to complete the task, users would have to split their attention between screen and input device. Probably, non-expert users fixed their attention only to the remote control because they were more focused on pressing the remote control's keys rather than on watching the TV screen output.

- **Localisation of the blank key**. All non-expert users needed help to find the blank key even if they were told that it was associated to the standard "0" key (like on the majority of mobile phones). Instead, expert users had no problems in finding it.

The virtual keyboard interface showed these problems:

- **Confirm the right character**. Even if during the training time users were told to press the "OK" key to confirm the selection, both expert and non-expert users frequently forgot to press it.

- **Slowness of text editing and frustration**. The average time spent in completing both tasks using the virtual keyboard interface was higher than the average time spent in completing the tasks using the other two interfaces (see Table 1). Even if at the beginning users were enthusiastic of the easiness of interaction, later, most of them became frustrated by the long time required to complete the tasks.

- **Localisation of the blank button**. The virtual keyboard interface layout (see Figure 5) follows the typewriter metaphor, therefore the blank button is represented by a bar button without any label. Most of the non-expert users (60%) did not understand the metaphor and, consequently, did not find the blank button.

Considering the difficulties expressed by the users in the questionnaire, we did not find meaningful differences in problems underlined both by expert and non-expert users: (1) multi-press with timeout ($X^2 = 4.267$, $df = 5$, $N = 36$, $p = 0.512$); (2) multi-press with timeout with visual feedback ($X^2 = 5.026$, $df = 5$, $N = 36$, $p = 0.413$); and (3) virtual keyboard interfaces ($X^2 = 8.356$, $df = 5$, $N = 36$, $p = 0.138$).

Besides, we analysed the performance results of each interface, focusing on efficiency, effectiveness, and users' satisfaction.

Regarding the *efficiency*, we calculated the average time spent to complete each task, the percentage of completed tasks, and the average number of characters typed per second (cps). Examining the average time per task (see Table 1), we found that the multipress with timeout interface got the best performance (almost half the time of virtual keyboard).

Table 1. Average time per task in seconds

		Expert	Non-expert	Total
Multi-press	T. 1	55.45	151.4	93.8
	T. 2	23.2	75.9	48.8
Multi-press with visual feedback	T. 1	93	173.4	119.8
	T. 2	42.6	85	66.3
Virtual keyboard	T. 1	106.6	187.8	131.6
	T. 2	63.9	156.3	104

The percentage of non-completed tasks was lower in multi-press with timeout interface than in virtual keyboard and in multipress with timeout and visual feedback interfaces (see Table 2). In particular, the percentage of non-completed tasks (in all interfaces) for naive users is higher than the expert users' one. Notice that, due to the different number of completed tasks between the two subgroups, we could not make an *analysis of variance* (ANOVA) between expert and naive users' performance.

Finally, we considered the average cps. Multi-press with timeout interface was the highest performing again (see Table 3). Both for Task 1 (see Figure 8) and Task 2 (see Figure 9); the gap between expert users' cps and non-expert users' cps is very wide. Instead, we could not notice any significant gap between expert and non-expert users' cps in the other two interfaces.

The results of our studies can be compared, with some caution, to the results of previous researches. For example, the TNT method (see *Background* section) has better performance regarding cps (between 0.77 cps and 1.47 cps) but it is not realistic. In fact, this method recognises users' mistakes and does not move forward in the text since the correct letter has not been entered. In this way the users had no need to look for the cancel button on the remote control to correct the wrong character. Therefore, the expert performance in our multi-press interface is comparable to the

Table 2. Percentage of non-completed task

		Expert	Non-expert	Total
Multi-press	T. 1	0%	33.34%	16.6%
	T. 2	0%	5.5%	2.78%
Multi-press with visual feedback	T. 1	0%	50%	25%
	T. 2	0%	5.5%	2.78 %
Virtual keyboard	T.1	0%	55.5%	27.7%
	T.2	5.5%	27.78%	16.67%

Table 3. Average character per second (cps)

		Expert	Non-expert	Total
Multi-press	T. 1	0.29	0.17	0.15
	T. 2	0.48	0.26	0.17
Multi-press with feedback	T. 1	0.10	0.09	0.08
	T. 2	0.14	0.13	0.07
Virtual keyboard	T. 1	0.17	0.13	0.12
	T. 2	0.22	0.16	0.1

Figure 8. Average character per second (cps) in task 1

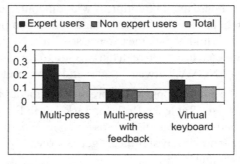

Figure 9. Average character per second (cps) in task 2

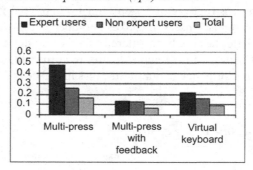

Table 4. Average errors per task

		Expert	Non-expert	Total
Multi-press	T. 1	3	9	5.27
	T. 2	1	2	2
Multi-press with visual feedback	T. 1	7	16	10
	T. 2	4	6	5
Virtual keyboard	T. 1	1	2	1
	T. 2	0	2	1

TNT one. We can also compare our study to the Graffiti and Jot input technique (Sears & Arora, 2002). Text entry speed with this method is in the range of 0.36-1.73 cps, comparable to the best performance of our multi-press interface. Finally, the estimated speed of multipress on a mobile phone (1.73-2.04 cps) is meaningfully higher than on a DTT handheld. However, this result can be explained considering DTT latency time.

Table 5. Cross-table: Subjective ratings for each interface and users' experience level

	Expert	Non-expert	Total
Multi-press	83.33	38.89	61.11
Multi-press with visual feedback	11.11	5.56	8.33
Virtual keyboard	0	22.22	11.11
All	5.56	27.77	16.67
No one	0	5.56	2.78

Summing up, considering average time, percentage of uncompleted tasks, and cps, multi-press interface scored the best result in terms of efficiency.

Regarding *effectiveness*, we analysed the average number of mistakes made during the text editing. Both expert and non-expert users made less mistakes using the virtual keyboard than using the other two interfaces (see Table 4). This is due to the features of this interface: All symbols and alpha-numeric characters are shown and users do not have to cycle through the letters.

Figure 10. Subjective ratings for each interface and users' experience level

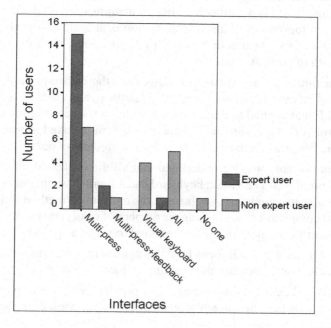

Questionnaires data were analysed to determine *users' satisfaction*. A meaningful relationship emerged between users' level of experience and favourite interface (X^2 = 10.909, df = 4, N = 36, p = 0.028). Both expert and non-expert users rated the multi-press with timeout interface as their favourite one (see Table 5). Making a subgroup analysis (see Figure 10), we found that 83.33% of the expert users preferred the multi-press interface because it was the fastest one and the most similar to mobile phones text entry methods. Instead, just 38.89% of non-expert users chose the multi-press with timeout interface, while 22.22% preferred the virtual keyboard, and 27.77% liked all of them.

It is reasonable to suppose that non-expert users encountered problems with all the interfaces but they did not choose the "No one" answer because of the so-called "social desiderability" (Roccato, 2003).

Future Trends

As we said in the introduction, the future DTT scenario will regard interactive services for citizens. For example, users will be able to vote, to book a medical examination, to pay bills, and so on. That means that users will access and manage personal information and sensitive personal data (racial or ethnic origin, political opinions, physical or mental health or condition, sexual life, etc.) after some kind of login procedure. Users can login to an application using a smart card or a traditional login form. A smart card is a card containing an identification code. The user can insert it in the set-top-box in order to identify him or herself. Currently, because of the limited use of smart cards in Italian DTT applications, users can only login entering text with the remote control.

Considering the future scenario we described, we think that the study, the design, and the evaluation of innovative password-entering interfaces and methods is a relevant issue in the HCI field applied to the DTT context. In fact, since entering text using a remote control is still a problematic issue (as we demonstrated in this chapter), we can imagine that it would be harder to enter a starred password text.

In particular, in our analysis, we underlined that all the evaluated interfaces had some usability problems. The virtual keyboard had a smaller number of errors but a higher average time to complete a task. Instead, the multi-press method had a smaller average time to complete a task but a higher number of errors (this is probably due to the doubt about the number of presses required to select a special character).

Therefore, our further work will focus both on improving current non-starred text-entry interfaces and on evaluating the usability of password entry.

Regarding password entry, besides the usability problem, we have also to consider the privacy issue. In fact, because of the slowness of the virtual keyboard input method and the visibility on the TV screen of the selected characters, this kind of interface does not guarantee user's privacy. On the other hand, because of each remote control's key maps (at least four alpha-numeric characters) and because password's characters are starred, multi-press input method guarantees a higher level of privacy but less awareness of the selected character. In conclusion, we think it is important to give users some kind of feedback regarding the selected characters while keeping a high level of privacy. In other words, we aim to find an interface that balances the trade-off between privacy and usability.

Conclusion

In this chapter we compared three different interfaces for text entry using a DTT remote control. Results showed that all the interfaces had some usability problems. Despite considering both expert and non-expert users, multi-press interface emerged as the best one because of (1) the less average time per task, (2) the higher user satisfaction (see Figure 10), and (3) the higher cps.

Moreover, a meaningful relationship was found between the users' level of experience in entering text on mobile phones and the interface the user indicated as their favourite one. Instead, we did not notice a significant relationship between users' level of experience and type of editing problems expressed in the questionnaires.

Comparing our study to some previous analysis on other input methods, the TNT method had better performance regarding cps but it is not realistic. We also compared our study to the Graffiti and Jot input technique (Sears & Arora, 2002), and their text entry speed is comparable to the best performance of our multi-press interface. Finally, it is worthy pointing out how the estimated speed of multi-press on a mobile phone is significantly higher than on a DTT remote control.

The results of our research can be explained considering DTT limitations:

- **The mapping between letters and keys.** Each key is mapped to at least three letters and currently there is not a standard for remote control numeric keypads.

- **The focus of user's attention**. In a DTT context the user has to pay attention both to the screen and to the remote control.

- **The latency time**. Users have to wait a long time to see on screen the result of their input.

Despite DTT limitations, the multi-press with timeout interface had the best performance regarding efficiency and user's satisfaction. We are aware that we cannot consider the multi-press with timeout interface as a final solution to text input on a DTT remote control. In fact, we guess that it will not be suitable, for example, when text is starred (editing a password).

Acknowledgment

Special thanks to Gian Luca Matteucci and Ferdinando Ricchiuti for the support and management of the project. We are very grateful to Roberto Borri and Luca Broglio for their fruitful suggestions. We also thank the CSP staff for their participation in the project; especially we thank Diego Campisi, Roberto Politi, and the DTT-Team. Finally, we thank Carlotta for her special performance in the test session, despite that she is a granny of 80 years old.

References

Butts, L., & Cockburn, A. (2002). *An evaluation of mobile phone text input methods*. Paper presented at the Third Australasian Conference on User interfaces, Melbourne, Victoria, Australia.

DGTVi. (2004). *D-book*. Retrieved November 23, 2005, from http://www.dgtvi. it/stat/Allegati/D-BOOK%20V.1.pdf

European Telecommunications Standards Institute (ETSI). (2003). *Human factors (HF); User interfaces; Character repertoires, ordering rules and assignments to the 12-key telephone keypad (ES 202 130 V1.1.1)*. Retrieved from http://portal.etsi.org/docbox/EC_Files/EC_Files/es_202130v010101p.pdf

Ingmarsson, M., Dinka, D., & Zhai, S. (2004). *A numeric keypad based text input method*. Paper presented at the SIGCHI conference on Human factors in computing systems, Vienna, Austria.

Karat, C. M., Halverson, C., Horn, D., & Karat, J. (1999). Patterns of entry and correction in large vocabulary continuous speech recognition systems. In ACM Press (Ed.), *ACM Conference on Human Factors in Computing Systems* (pp. 568-575). New York: ACM Press.

MacKenzie, I. S. E. (2002). Special issues on text entry for mobile devices. *Human-Computer Interaction, 17*(2,3).

Miller, R.M. (1968). *Response time in mancoputer conversational transactions*. AFIPS Conference Proceedings (pp. 267-277).

Nielsen, J., (n.d.). *Useit.com: Jakob Nielsen's website.* Retrieved from http://www. useit.com/

Norman, D. A. (1988). *The psychology of everyday things.* New York: Basic Books.

Pavlovych, A., & Stuerzlinger, W. (2004). *Model for non-expert text entry speed on 12-button phone keypads.* Paper presented at the SIGCHI conference on Human factors in computing systems, Vienna, Austria. ACM Press.

Roccato, M. (2003). Desiderabilità sociale e acquiescenza. Alcune trappole delle inchieste e dei sondaggi. LED Edizioni Universitarie.

Sears, A., & Arora, R. (2002). Data entry for mobile devices: An empirical comparison of novice performance with Jot and Graffiti. *Interacting with Computers, 14*(5), 413-433.

Silfverberg, M., MacKenzie, I. S., & Korhonen, P. (2000, April). *Predicting text entry speed on mobile phones.* Paper presented at the Conference on Human Factors in Computing Systems, The Hague, The Netherlands. ACM Press.

Zhai, S., Smith, B. A., & Hunter, M. (2003). Performance optimization of virtual keyboarding. *Human-Computer Interaction, 17*(2,3), 89-129.

Chapter XIV

Getting the Big Picture on Small Screens:
Quality of Experience in Mobile TV

Hendrik Knoche, University of College London, UK

M. Angela Sasse, University of College London, UK

Abstract

This chapter provides an overview of the key factors that influence the quality of experience (QoE) of mobile TV services. It compiles the current knowledge from empirical studies and recommendations on four key requirements for the uptake of mobile TV services: (1) handset usability and its acceptance by the user, (2) the technical performance and reliability of the service, (3) the usability of the mobile TV service (depending on the delivery of content), and (4) the satisfaction with the content. It illustrates a number of factors that contribute to these requirements ranging from the context of use to the size of the display and the displayed content. The chapter highlights the interdependencies between these factors during the delivery of content in mobile TV services to a heterogeneous set of low resolution devices.

Introduction

It is the second time around for mobile TV. In the 1980s, Seiko introduced a TV wristwatch that was capable of displaying standard TV channels on an liquid crystal display (LCD) wrist watch. It seemed like a great idea at the time. Many people wore watches, a growing number of people used LCD or digital watches, and it was possible to display anything on an LCD display. However, the watch was not a success. One of the biggest problems was high energy consumption—the watch wearer had to separately carry the battery, which was part of a box that housed the TV receiver and connected to the watch through a cable. This setup gave the wearer approximately one hour of viewing time. The screen was monochrome and had low contrast. Furthermore, watching TV while wearing the watch resulted in an unnatural wrist posture. Last but not least, the TV wristwatch was expensive.

Twenty years later, mobile TV is back. Many people now carry inexpensive mobile phones with built-in LCD screens. This allows the display of moving images, which can be received in a more energy efficient way these days, and mobile TV is making its second appearance. Today, mobile TV services are available in a number of countries. While Asian consumers already have access to broadcast services, Western countries have finished trials and are aiming to move from unicast, that is, individual delivery services, to broadcast solutions. Portable play stations and video Ipods provide alternative platforms for playing prestored content.

So far, the deployment of these services has been driven by technical feasibility and matching business models. The wireless domain is one of limited bandwidth resources, and service providers have to decide on broadcasting more content at lower quality or vice versa in search of optimal configurations for people's QoE that are financially viable. The content is produced by companies with a specific primary target medium, that is, cinema, TV, or mobile in mind. This choice influences the selection of shot types, length, and the type of programme. Cameras can be chosen from a wide selection delivering different resolutions, aspect ratios, contrast ranges, and frame rates. After post-production the content is delivered to audiences through various channels. For example, TV broadcast companies adapt cinema content to the TV and mobile service operators adapt TV content for mobile TV distribution. Uptake of existing mobile TV services lags behind expectations, possibly because customers are not willing to pay high premiums for content (KPMG, 2006). To assist service providers in improving their service offerings, we need to understand how people might experience mobile TV services in their entirety. QoE (Aldrich, Marks, Lewis, & Seybold, 2000; Jain, 2004; McCarthy & Wright, 2004) is a broad concept that encompasses all aspects of a service that can be experienced by the user. In the case of mobile TV, QoE includes the usability of the service; the restrictions inherent in the delivery; the audio-visual quality of the content; the usage and payment model; and the social context as well as possible parallel use of standard TV.

According to Mäki (2005) the following four requirements are the most important for adoption of mobile TV services:

1.   Handset usability and acceptance
2.   Technical performance and reliability
3.   Usability of the mobile TV service
4.   Satisfaction with the content

We will address each of these factors in turn in more detail in the following sections in order to provide a comprehensive view on the QoE of mobile TV services.

Handset Integration and Usability

Currently, the mobile phone is the most likely platform for mobile TV, but personal digital assistants (PDAs), portable game consoles, and music players are attractive alternatives. In 2003 a total of 70% of the people in Europe owned or used mobile phones. The importance of mobile phones in people's lives means that most owners carry it with them wherever they go. Mobile TV consumption on mobile phones allows for privacy of consumption, because of short viewing distances and the viewing angle afforded by many mobile devices. However, people perceive the battery consumption of mobile TV as a threat to more important communication needs. The application should provide warnings when the battery is drained beyond a certain threshold (Knoche & McCarthy, 2004), and service providers should set user expectations about battery drain induced by, for example, watching live content (Serco, 2006).

The mobile TV application should not get in the way of communication but alert users of incoming calls, text, or other messaging and provide means to deal with them in a seamless manner. On inbound communication this includes automatic pausing of the TV service, if possible, and offering to resume once the user has finished communicating. Likewise, important indicators, for example, for battery status or menus should not unnecessarily obstruct the TV screen but could use semi-transparent menus as suggested in Serco (2006).

Depending on the technical realisation, having TV reception might require a second receiver unit in the handset that would allow for parallel reception of TV content and making and receiving telephone calls at the same time. A single receiver unit, for example, would not be able to record live TV content during a phone call.

Display

The screen should have high contrast, backlight, and a high viewing angle to support viewing in different circumstances and by multiple viewers. Due to size and power constraints LCDs are currently the preferred technology to present visual information on mobile devices. LCDs come in a range of shapes, sizes, and resolutions, from video graphics array (VGA) PDAs (480 x 640 pixels) and high end third generation (3G) or digital video broadcasting-handheld (DVB-H) enabled phones (320 x 240) to more compact models with quarter common intermediate format (QCIF) size (176 x 144) and below. Users want as large a screen as possible for viewing, but they do not want their phones to be too big (Knoche & McCarthy, 2004). Landscape-oriented use of the display might be preferred (Serco, 2006) over the typical portrait mode that mobile phones are used in. In general, pictures subtending a larger visual angle in the eye of the beholder make for a better viewing experience. Results from studies on TV pictures revealed that larger image sizes are generally preferred to smaller ones (Lombard, Grabe, Reich, Campanella, & Ditton, 1996; Reeves & Nass, 1998) and are perceived to be of higher quality (Westerink & Roufs, 1989), but that there is no difference in arousal and attention between users watching content on 2" and 13" screens (Reeves, Lang, Kim, & Tartar, 1999).

Which resolutions best support the different screen sizes of mobile TV devices is subject to current research, and the pros and cons of different resolutions will be discussed in the section on video quality. Besides the size of the device, the visual impact can be increased by head-mounted displays and projection techniques. Whereas, the former results in an experience of greater immersion and requires additional equipment to be carried around, the latter reduces the anonymity of visual consumption.

Device Use

A stand for continuous viewing is beneficial for mid- to long-term use. In public places, the use of headphones, which increase immersion (see later on), might be required. Many people already use head phones for portable music players, and standard headset jacks on mobile TV devices would make switching between devices easy. Dedicated buttons would be valuable for mobile TV access, basic playback controls, and content browsing, for example, channel switching or selecting. On touch screens many people value on-screen buttons that do not decrease the viewing area of the content instead of having to use a stylus which requires two-handed operation.

Immersion

Users are worried about becoming too absorbed in what they are watching and thus distracted from other tasks while being on the move, for example, missing trains or stops (Knoche & McCarthy, 2004). They require a pause/mute facility to cope with likely interruptions. In the case of broadcast content, this requirement places demands on the device's storage capacity. Volume control should possible preferably without the need to access menus. The question whether a separate means to mute the volume and would let the video play in the background will be necessary or might confuse users more in conjunction with the pause button which pauses both audio and video has to be addressed by future research. An easy way to set alarms or countdowns might help mobile users to not loose touch with the world around them.

Technical Performance and Reliability

The way the content is delivered has a major effect on the possible uses of a mobile TV service. The perceived video and audio quality will depend on the quality of service (QoS) provided by the network that is delivering the packets that carry the content and might noticeably degrade the content by introducing errors, loss, and varying delays to those packets. For an example of how loss influences the perceived video quality of mobile TV content see Jumisko-Pyykkö, Vinod Kumar, Liinasuo, and Hannuksela (2006).

Service Delivery

From the user point of view, TV is commonly understood as an "any time" service: turn it on and it will deliver content at any time of day. Mobile TV services implicitly suggest being available anywhere at any time. Mobile phone users have been wary of this promise (Knoche & McCarthy, 2004).

There are four content delivery models that significantly shape the experience of the mobile TV service: (1) media charger, (2) streaming (unicast), (3) broadcast, and (4) pre-cached broad- or multi-cast.

The video Ipod is an example of a *media charger*. The user has no live content but does have full playback control and can watch anywhere at any time. In order to have a supply of fresh content the user has to touch base regularly.

Many of the services like MobiTV and Slingbox (Sling Media, 2006) in the U.S. and Vodafone live! in the UK are currently offered as unicast services, which makes them relatively expensive in terms of spectrum usage and difficult to scale. With each increase in the number of receivers in a reception cell the available bandwidth per receiver decreases. The number of users receiving a unicast mobile TV stream on demand within a wireless cell is therefore limited, and the audio-visual quality degrades with the increase in receivers. However, unicast can deliver personalised content for niche interests that would not be viable through broadcasts.

Broadcast approaches like digital multimedia broadcasting (DMB) and DVB-H are more efficient in mass delivery as they support an arbitrary number of receivers at constant quality in the coverage area. Broadcast users have no playback control unless pausing live TV and other functions available in personal video recorders (PVR) are implemented on the user terminal. However, since being on the move results in varying levels of reception people experience varying quality and service discontinuities. This poses a problem to broadcast TV services without PVR-like functionality. People might tune into the streams at times when the programmes they want to see are not being broadcast. Similar to media chargers, pre-cached services, for example, SDMB (Selier & Chuberre, 2005), can continuously display recently downloaded, that is, non-live, content at higher quality through carrousel broadcasts. This is an example of TV any time, which allows users to watch broadcast content when convenient. Mobile TV services do not have to rely solely on one of these delivery mechanisms but could mix them in order to leverage their different advantages.

The content has to be delivered through one of these transmission schemes to a range of devices with different display capabilities. There are three main ways to address the problem of multiple target solutions: (1) sending multiple resolutions, which requires more bandwidth if broadcast; (2) broadcasting at the highest resolution and resizing at the receiver side; and (3) employing layered coding schemes that broadcast a number of resolution layers from which every receiver can assemble the parts it can display.

Resolution, Image Size, and Viewing Distance

Human perception of displayed information has been studied for a long time, see Biberman (1973) for an overview. Spatial and temporal resolution are key factors for the perceived quality of video content. Whereas, temporal resolution below 30 frames per second (fps) results in successively jerkier motion, lowering the number of pixels to encode the picture reduces the amount of visible detail. Excessive delays and loss during transmission of the content may affect both the spatial and temporal resolution resulting in visible artefacts and or skipping of frames causing the picture to freeze.

The higher the resolution in both of these dimensions, the more bandwidth is required to transmit it. Service providers only have a limited amount of bandwidth available and want to maximise the content they can offer to their customers while still delivering the quality that the customers expect. They face the trade-off between visual quality and quantity of the content.

Mobile TV will be consumed at arm's length. Paper, keyboard, and display objects are typically operated at distances ranging from 30 cm to 70 cm. Continued viewing at distances closer than the resting point of vergence—approx. 89 cm, with a 30° downward gaze—can contribute to eyestrain (Owens & Wolfe-Kelly, 1987). When viewing distances come close to 15 cm, people experience discomfort (Ankrum, 1996). Normal 20/20 vision is classified as the ability to resolve 1 minute of arc (1/60°) (Luther, 1996) and translates to 60 pixels per degree. The amount of pixels p that can be resolved by a human at a given distance d and a picture height h can be computed by the following equation: $p = \dfrac{h}{d \cdot 2\tan(1/120)}$.

In the typical TV viewing setup at a seating distance of 3m, the benefits of high definition television (HDTV) can only be enjoyed on relatively big screens. On handheld devices, people could easily enjoy HDTV resolutions on a screen of 8 cm height. However, mobile TV does not exceed quarter video graphics array (QVGA) resolution at present. In addition, people are able to identify content that has been upscaled from low resolutions to higher resolution mobile screens. So far no research has addressed the potential effects of upscaling low broadcast resolution content to a screen with a higher resolution. Research on these topics is proprietary. Philips uses a nonlinear upscaling method called Mobile PixelPlus to fill a screen with higher resolution than the broadcast material.

Some studies have addressed the perception of low resolution content on small handheld screens (Knoche, McCarthy, & Sasse, 2005; Jumisko-Pyykkö & Häkkinen, 2006; Song, Won, & Song, 2004). Content shown on mobile devices at higher resolutions and larger sizes is generally more acceptable than lower resolutions and smaller sizes at identical encoding bitrates. However, the differences are not uniform across content types (Knoche, McCarthy, et al., 2005). All content types received poor ratings when presented at resolutions smaller than 168 x 126. Other studies have shown that low image resolution can improve task performance. For example, Horn (2002) showed that lie detection was better with a small (53 x 40) than a medium (106 x 80) video image resolution. In another study, however, smaller video resolutions (160 x 120) had no effect on task performance but did reduce satisfaction when compared to 320 x 240 image resolutions (Kies, Williges, & Rosson, 1996). In a study by Barber and Laws (1994), a reduction in image resolution (from 256 x 256 to 128 x 128) at constant image size led to a loss in accuracy of emotion detection especially in a full body view. The legibility of text has a major influence on the acceptability of the overall video quality (Knoche & Sasse, 2006) and should be sent separately and rendered at the receiving side.

Frame Rate

Low video frame rates are common in recent mobile multimedia services especially in streamed unicast services. Frame rates as low as 5 fps and lower were avoided at all costs in a desktop computer-based study by Pappas and Hinds (1995). Another study, conducted by Apteker, Fisher, Kisimov, & Neishlos (1994) assessed the watchability of various types of video at different frame rates (30, 15, 10, 5 fps). Compared to a benchmark of 100% at 30 fps, video clips high in visual importance dropped to a range of 43% to 64% watchability when displayed at 5 fps, depending upon the importance of audio for the comprehension of the content and the static/dynamic nature of the video. Participants who saw football clips on mobile devices found the video quality of football content less acceptable when the frame rate dropped below 12 fps (McCarthy, Sasse, & Miras, 2004). Comparable displays on desktop computers maintained high acceptability for frame rates as low as 6 fps. The reason for the higher sensitivity to low frame rates on mobile devices is not yet fully understood, but highlights the importance to measure video quality in as realistic setups as possible to the real experience. The proprietary natural motion approach by Philips supposedly reduces the jerkiness of low-frame-rate content by generating intermediate frames from the broadcast set of frames at the receiver side (De Vries, 2006).

Some programmes have sign language interpreters signing to make the programme understandable for deaf people. This is one of the few applications that require high frame rates for comprehension of the visual content. Spelling sign language requires 25 frames to be able to capture all letters in at least one frame (Hellström, 1997).

Temporal vs. Spatial Resolution

Whereas, earlier guidelines suggested the use of higher frame rates for fast moving content, for example, sports, (IBM, 2002) recent findings show that users prefer higher spatial resolution over higher frame rates in order to be able to identify objects and actors in mobile TV content (McCarthy et al., 2004). Wang, Speranza, Vincent, Martin, and Blanchfield (2003) reported on a study in which they manipulated both frame rate and quantization with an American football clip. They concluded that "quantization distortion is generally more objectionable than motion judder" and that large quantization parameters should be avoided whenever possible.

Audio Visual Quality

A number of studies have found that the combined quality of audio-visual displays is not simply based on the sum of its parts (e.g., Hands, 2004; Jumisko-Pyykkö et et al., 2006). In a study on audio-visual interactions, Winkler and Faller (2005) found that selecting mono audio for a given bitrate gives better quality ratings and that more bitrate should be allocated to the audio for more complex scenes. As a byproduct in a study on TV viewing experience Neuman, Crigler, and Bove (1991) discovered that the perceived video quality was improved by better audio. However, it was only the case for one of the three used content types. Similarly, a study by Beerends and De Caluwe (1999) using a 29 cm monitor, found that the rating of video quality was slightly higher when accompanied by CD quality audio than when accompanied by no audio. The effect, however, was small and has not been replicated with small screens. However, in the same study participants judged the two lower video quality levels (in which the video bandwidth was limited to 0.15 MHz and 0.025 MHz) worse when they were presented with audio, than without audio. Similarly, in a study by Knoche, McCarthy, et al. (2005), the visual quality of video clips displayed on mobile devices was more acceptable to participants across all video encoding bitrates when it was supported by lower (16 kbps) than with higher audio quality (32 kbps).

Synchronous playback of sound and video affects the overall AV-quality (Knoche, De Meer, & Kirsh, 2005). For 30 fps video the window of synchronisation is ±80ms (Steinmetz, 1996). The temporal window of synchronisation depends on the video frame rate (Knoche, De Meer, et al., 2005; Vatakis & Spence, in press). At lower frame rates audio-visual speech perception is more sensitive to audio coming before video and the presentation of the audio relative to the video should be delayed (Knoche, De Meer, et al., 2005).

Usability of the Mobile TV Service

In order to understand what makes for a usable mobile TV service, we need to know about the context of the user including the motivation for use and the location. Many of the guidelines that apply to mobile application design in general equally apply to mobile TV; see Serco (2006) for an overview.

Motivation of Use

Whereas, the drivers behind standard TV consumption are fairly well understood, we lack comparable knowledge in mobile TV. Peoples' watching of standard TV is driven by ritualistic (Taylor & Harper, 2002) and instrumental motives (Rubin, 1981) as in "electronic wallpaper" (Gauntlett & Hill, 1999), mood management (Zillman, 1988), escapism, information, entertainment, social grease, social activity, and social learning (Lee & Lee, 1995). For many of these drivers watching TV constitutes a group activity. Mobile TV is, due to its nature and limitations, more likely to be an individual consumption activity. The restricted viewing angle of the screens, the (for some people uncomfortable) proximity with others to share it, and the fact that the mobile phone is a rather personal device might curb group usage.

Location

According to Mäki (2005) the most common places for mobile TV use are (in descending order):

1.   in public transport
2.   At home
3.   At work

This is supported by other studies, in which many participants of mobile TV trials used the device as an additional TV set at home (Södergård, 2003). While at home, users' perception of the mobile TV service might depend on the comparison with standard TV in terms of delay (mobile broadcasts might incur additional delay due to processing or delivery, e.g., through satellite); programme availability; audio-visual quality; responsiveness; ease of use; interoperability with other media solutions including recording devices—such as PVRs—that allow for easy recording of television shows; content sharing; and user-controlled storage.

People are able to compare the different experiences of consuming TV content at home. Some might object to the inherent delay (approximately 1 minute) between the live broadcast TV signal and the mobile TV signal as currently seen in MobiTV (Lemay-Yates Associates Inc., 2005). What is more important, perhaps, is that the delay disadvantages the mobile audience in interactive game shows or betting services.

Usage Patterns

Previous research has shown that peoples' average usage of mobile TV is less than 10 minutes long (Södergård, 2003). This window of consumption places demands both on the length of consumable content and the time that users might be willing to spend to access and navigate through it. Data from SDMB trials in Korea for example show that people use mobile TV throughout the day with peaks in the morning, at lunch time, in the early evening, and very late in the evening.

Interactivity of the Mobile TV Service

The interface needs to provide the user with controls to use the different kinds of interactivities offered in mobile TV. Users expect the entry points to the mobile TV to be available from prominent places in the mobile phone user interface (Stockbridge, 2006).

Participation Interactivity and Payments

One of the potentially biggest advantages of mobile TV over regular TV is the existence of a return channel with built-in billing possibilities for premium and subscription services, as well as transactions involved in interactive services such as voting and betting.

Participants in mobile TV trials favoured the flat-rate payment model, that is, a single payment for unlimited mobile TV use during a billing period (Mäki, 2005). Flat rates do not place additional barriers between the users and the content. In South Korea, early payment models greatly influenced the use of mobile TV. When mobile TV usage was billed in the amount of kilobytes received, each 1-minute part of a programme made especially for mobile TV had to be confirmed for delivery, which resulted in a discontinuous viewing experience (Knoche, 2005).

Distribution Interactivity and Content Navigation

Taylor and Harper (2002) argued that channel surfing is inherently associated with the act of watching TV. The methods to select a programme used in traditional TV viewing depend on the time of day. But the method used generally escalates—if nothing of interest is found—to strategies that require more effort on behalf of the user. The order of strategies is:

1.   Channel surfing
2.   Wait or search for a TV programme announcement
3.   Knowledge of weekly schedules or upcoming programmes
4.   Paper-based or on-screen guides

Since mobile TV usage spurts are rather short, waiting for and searching for announcements or upcoming programmes might not be feasible. Information on what is currently playing and what will come up next might be valuable and should be easy to access.

Ideally, dedicated buttons or soft keys will allow users to switch channels. Long waiting times after a requested channel switch will result in lower user satisfaction. Tolerable switching delays between mobile TV channels have not been thoroughly researched but should be as short as possible since users are accustomed to almost instantaneous switches on standard TV. First results for digital TV indicate that 0.43 seconds might be the limit beyond which users will be increasingly dissatisfied (Ahmed, Kooij, & Brunnström, 2006). In digital TV, the switching delays depend to a large part on the video codec, for example, in Motion Picture Experts Group (MPEG)-encoded content on the occurrence of so-called key frames. Fewer key frames in a video broadcast result in smaller amounts of bandwidth required to transmit the content, but the receiver has to wait for the arrival of the next key frame in order to be able to display a newly selected channel. Service providers could exploit the fact that the human visual system is inert. An average recovery time of 780 msec between scene changes was acceptable to even the most critical observers, when visual detail was reduced to a fraction of the regular stream (Seyler & Budrikis, 1964). Further research would be needed to see if this period applies equally to channel switching on mobile devices and how which codecs could make use of this period. Displaying the logo of the upcoming channel or other tricks might perceptually shorten the wait time for users. Long wait times, for example, for downloading or on-demand streaming content should be accompanied with progress bars to help users assess the remaining time (Serco, 2006).

Because of the strong brand recognition of current TV broadcasters (e.g., CNN, BBC), it is likely that channel-centric content organisation under those brand names will prevail in mobile TV. But they could be replaced by virtual channels (Chorianopoulos, 2004) or category-centric content organisation which would group similar content from various sources under one category (e.g., news, music, movies, etc.). Because of the limited space and need for fast access, users will be interested in arranging content and channels according to their preferences. An electronic programme guide (EPG) which shows what programmes are currently available for viewing and what will come up might become a more important content navigation tool in mobile TV than in digital TV settings as reported in Eronen and Vuorimaa (2000).

Different video skipping approaches (Drucker, Glatzer, De Mar, & Wong, 2004), skimming video (Chistel, Smith, Taylor, & Winkler, 2004), and overall gist determination and information seeking (Tse, Vegh, Marchionini, & Shneiderman, 1999) have been studied in digital and standard TV settings but not in the mobile domain. When selecting from a range of programmes represented by video clips playing in parallel on mosaic pages of a digital TV study found that viewers preferred interfaces that gave fewer choices and bigger pictures (Kunert & Krömker, 2006). This would have to be traded off with the necessary navigation required between pages or scrolling in order to display all possible channels of a big bouquet. Mobile TV services, which provide a mixture of live, pre-cached, and downloadable content need to communicate these differences through the user interface.

Information Interactivity

Accessing additional information on mobile TV programmes is a challenge to design because of the limited screen estate. While watching regular TV some people are already making use of their mobile phones by sending short message service (SMS) messages to friends to comment on what they are watching on TV. This kind of distributed co-viewing experience would be feasible on mobile TVs with large enough screens to show both the content and the textual conversation.

Digital Rights Management (DRM)

People have a strong sense of ownership about the content that resides on their mobile devices. Many expect to be able to capture and transfer the content to and from computers for back-up purposes or for sharing with friends (Knoche & McCarthy, 2004). Restrictive DRM approaches that run against perceived user needs will affect the experience of mobile TV.

Content

The content distributed to mobile devices ranges from interactive content, specifically created for the mobile, to material that is produced for standard TV or cinema consumption. A number of studies have identified news as the most interesting content for mobile consumption(Knoche & McCarthy, 2004; Mäki, 2005). Considering the fact that many users watch mobile TV at home there is not much reason why programmes on regular TV would not be popular on mobile devices unless they prove impractical to watch on small screens. Whereas, news is of interest throughout the day, participants want to watch sports; series and general entertainment; music; and films on specific occasions (Mäki, 2005). Many people expect that their standard TV channels will be available on mobile TV (Serco, 2006). Time will tell whether relaying standard TV channels will be good enough for a mobile audience that is constrained when to watch, has short viewing periods, and small display sizes.

Made for Mobile Content

Currently, content made especially for mobile use is expensive as the audience compared to broadcast television is relatively small. However, content producers adapt their content with respect to low resolutions and the typical use time, for example, short versions of the popular TV series 24. In sports coverage for mobile devices ESPN minimises the use of long shots in their coverage (Gwinn & Hughlett, 2005) and instead uses more highlights with close-up shots. Others produce soap operas for mobile devices that rely heavily on close-up shots with little dialogue ("Romantic drama in China soap opera," 2005). However, the gain of these changes is not fully researched or understood. Research has shown that differences in the perceived quality of shot types depend on the displayed content (Knoche, McCarthy, & Sasse, 2006). Further research is required to evaluate the potential benefits of cropping for mobile TV resolutions.

Recoded Content

Relatively cheap in comparison to the made-for-mobile content is the pre-encoding of cinema or TV content both in length and in size. Automatic highlight extraction from TV content (Voldhaug, Johansen, & Perkis, 2005) is a promising technique that needs to be evaluated with end users on mobile devices.

Content-based pre-encoding can improve on the visual information and detail by: (1) cropping off the surrounding area of the footage that is outside the final safe area for action and titles and does not include essential information (Thompson, 1998); (2) zooming in on the area that displays the most important aspects (Dal Lago, 2006; Holmstrom, 2003); and (3) visually enhancing content, for example, by sharpening the colour of the ball in football content (Nemethova, Zahumensky, & Rupp, 2004). Research is required to rule out possible negative side effects caused by these automated approaches.

Future Trends

Video encoders will further reduce the amount of encoding bitrates required and will result in better perceived quality. Memory will continue to drop in price and make full PVR functionality with ample amounts of storage capacity available on mobile TV devices. Designing a mobile TV service on the edges of the coverage area might be another challenge. When viewers move in and out of the coverage area or the kind of delivery service that is provided the application will have to feature a way to gracefully switch between these different service concepts, for example, DVB-H live streams and pre-cached content in SDMB. Intelligent cropping algorithms that enlarge parts of the content might become a solution if the content depicted on mobile TV screens is too small for the viewer. Mobile phones with video camera capabilities might make for a very different mobile TV experience if peers or groups of people start providing each other with video clips on the go.

Conclusion

Mobile TV is a very promising service for both customers and service providers. In order to provide the former with a satisfying QoE during potentially short interaction periods, the service provider will have to take into consideration a range of aspects in the creation, preparation, delivery, and consumption of content on a variety of mobile platforms. It will require cooperation between all involved parties to make mobile TV as appealing as the standard TV that constitutes a necessity in many households. This chapter has presented the key factors that determine QoE for mobile TV along with previous research results which can help improve the uptake of mobile TV 2.0 in a mobile and diversified market place.

References

Ahmed, K., Kooij, R., & Brunnström, K. (2006). Perceived quality of channel zapping. In *ITU-T Workshop on QoE/QoS 2006*.

Aldrich, S. E., Marks, R. T., Lewis, J. M., & Seybold, P. B. (2000). *What kind of the total customer experience does your e-business deliver?* Patricia Seybold Group.

Ankrum, D. R. (1996). Viewing distance at computer workstations. *Work Place Ergonomics,* 10-12.

Apteker, R. T., Fisher, A. A., Kisimov, V. S., & Neishlos, H. (1994). Distributed multimedia: User perception and dynamic QoS. In *Proceedings of SPIE* (pp. 226-234).

Barber, P. J., & Laws, J. V. (1994). Image quality and video communication. In R. Damper, W. Hall, & J. Richards (Eds.), *Proceedings of IEEE International Symposium on Multimedia Technologies & their Future Applications* (pp. 163-178). London, UK: Pentech Press.

Beerends, J. G., & De Caluwe, F. E. (1999). The influence of video quality on perceived audio quality and vice versa. *Journal of the Audio Engineering Society, 47*, 355-362.

Biberman, L. M. (1973). *Perception of displayed information*. Plenum Press.

Chistel, M., Smith, M., Taylor, C., & Winkler, D. (2004). Evolving video skims into useful multimedia abstractions. In *Proceedings of CHI '98* ACM Press.

Chorianopoulos, K. (2004). *Virtual television channels conceptual model, user interface design and affective usability evaluation*. Unpublished doctoral thesis, Greece: Athens University of Economics and Business.

Dal Lago, G. (2006). *Microdisplay emotions*. Retrieved from http://www.srlabs. it/articoli_uk/ics.htm

De Vries, E. (2006). *Renowned Philips picture enhancement techniques will enable mobile devices to display high-quality TV images*. Retrieved from http://www. research.philips.com/technologies/display/picenhance/index.html

Drucker, P., Glatzer, A., De Mar, S., & Wong, C. (2004). SmartSkip: Consumer level browsing and skipping of digital video content. In *Proceedings of the SIGCHI conference on Human factors in computing systems: Changing our world, changing ourselves* (pp. 219-226). New York: ACM Press.

Eronen, L., & Vuorimaa, P. (2000). User interfaces for digital television: A navigator case study. In *Proceedings of the Working Conference on Advanced Visual Interfaces AVI 2000* (pp. 276-279). New York: ACM Press.

Gauntlett, D., & Hill, A. (1999). *TV living: Television, culture and everyday life*. Routledge.

Gwinn, E., & Hughlett, M. (2005, October 10). Mobile TV for your cell phone. *Chicago Tribune*. Retrieved from http://home.hamptonroads.com/stories/story. cfm?story=93423&ran=38197

Hands, D. S. (2004). A basic multimedia quality model. *IEEE Transactions on Multimedia, 6*, 806-816.

Hellström, G. (1997). Quality measurement on video communication for sign language. In *Proceedings of 16th International Symposium on Human Factor in Telecommunications* (pp. 217-224).

Holmstrom, D. (2003). *Content based pre-encoding video filter for mobile TV*. Unpublished thesis, Umea University, Sweden. Retrieved from http://exjob. interaktion.nu/files/id_examensarbete_5.pdf

Horn, D. B. (2002). The effects of spatial and temporal video distortion on lie detection performance. In *Proceedings of CHI' 02*.

IBM. (2002). *Functions of mobile multimedia QOS control*. Retrieved from http:// www.trl.ibm.com/projects/mmqos/system_e.htm

Jain, R. (2004). Quality of experience. *IEEE Multimedia, 11,* 95-96.

Jumisko-Pyykkö, S., & Häkkinen, J. (2006). "I would like see the face and at least hear the voice": Effects of screen size and audio-video bitrate ratio on perception of quality in mobile television. In G. Doukidis, K. Chorianopoulos, & G. Lekakos (Eds.), *Proceedings of EuroITV '06* (pp. 339-348). Athens: University of Economics and Business.

Jumisko-Pyykkö, S., Vinod Kumar, M. V., Liinasuo, M., & Hannuksela, M. (2006). Acceptance of audiovisual quality in erroneous television sequences over a DVB-H channel. In *Proceedings of the Second International Workshop in Video Processing and Quality Metrics for Consumer Electronics.*

Kies, J. K., Williges, R. C., & Rosson, M. B. (1996). *Controlled laboratory experimentation and field study evaluation of video conference for distance learning applications* (Rep. No. HCIL 96-02). Blacksburg: Virginia Tech.

Knoche, H. (2005). *First Year report.* Unpublished thesis, University College London.

Knoche, H., De Meer, H., & Kirsh, D. (2005). Compensating for low frame rates. In *CHI' 05 extended abstracts on Human factors in computing systems* (pp. 1553-1556).

Knoche, H., & McCarthy, J. (2004). Mobile users' needs and expectations of future multimedia services. In *Proceedings of the WWRF12.*

Knoche, H., McCarthy, J., & Sasse, M. A. (2005). Can small be beautiful? Assessing image resolution requirements for Mobile TV. In *ACM Multimedia* ACM.

Knoche, H., McCarthy, J., & Sasse, M. A. (2006). A close-up on mobile TV: The effect of low resolutions on shot types. In G. Doukidis, K. Chorianopoulos, & G. Lekakos (Eds.), *Proceedings of EuroITV '06* (pp. 359-367). Greece: Athens University of Economics and Business.

Knoche, H., & Sasse, M. A. (2006). Breaking the news on mobile TV: User requirements of a popular mobile content. In *Proceedings of IS&T/SPIE Symposium on Electronic Imaging.*

KPMG. (2006). Consumers and convergence challenges and opportunities in meeting next generation customer needs.

Kunert, T., & Krömker, H. (2006). Proven interaction design solutions for accessing and viewing interactive TV content items. In G. Doukidis, K. Chorianopoulos, & G. Lekakos (Eds.), *Proceedings of EuroITV 2006* (pp. 242-250). Greece: Athens University of Economics and Business.

Lemay-Yates Associates Inc. (2005). *Mobile TV technology discussion.* Lemay-Yates Associates Inc.

Lombard, M., Grabe, M. E., Reich, R. D., Campanella, C., & Ditton, T. B. (1996). Screen size and viewer responses to television: A review of research. In *Annual Conference of the Association for Education in Journalism and Mass Communication*.

Luther, A. C. (1996). *Principles of digital audio and video*. Boston, London: Artech House Publishers.

Mäki, J. (2005). *Finnish mobile TV pilot*. Research International Finland.

McCarthy, J., Sasse, M. A., & Miras, D. (2004). Sharp or smooth? Comparing the effects of quantization vs. frame rate for streamed video. In *Proceedings of CHI* (pp. 535-542).

McCarthy, J., & Wright, P. (2004). *Technology as experience*. Cambridge, MA: MIT Press.

Nemethova, O., Zahumensky, M., & Rupp, M. (2004). Preprocessing of ball game video-sequences for robust transmission over mobile networks. In *Proceedings of the CIC 2004 The 9th CDMA International Conference*.

Neumann, W. R., Crigler, A. N., & Bove, V. M. (1991). Television sound and viewer perceptions. In *Proceedings of the Joint IEEE/Audio Eng. Soc. Meetings* (pp. 101-104).

Owens, D. A., & Wolfe-Kelly, K. (1987). Near work, visual fatigue, and variations of oculomotor tonus. *Investigative Ophthalmology and Visual Science, 28*, 743-749.

Pappas, T., & Hinds, R. (1995). On video and audio integration for conferencing. In *Proceedings of SPIE—The International Society for Optical Engineering*.

Reeves, B., Lang, A., Kim, E., & Tartar, D. (1999). The effects of screen size and message content on attention and arousal. *Media Psychology, 1*, 49-68.

Reeves, B., & Nass, C. (1998). *The media equation: How people treat computers, television, and new media like real people and places*. University of Chicago Press.

Romantic drama in China soap opera only for mobile phones. (2005, June 28). *Guardian Newspapers Limited*. Retrieved from http://www.buzzle.com/editorials/6-28-2005-72274.asp

Rubin, A. M. (1981). An examination of television viewing motivations. *Communication Research, 9*, 141-165.

Selier, C., & Chuberre, N. (2005). Satellite digital multimedia broadcasting (SDMB) system presentation. In *Proceedings of 14th IST Mobile & Wireless Communications Summit*.

Serco. (2006). *Usability guidelines for Mobile TV design*. Retrieved from http://www.serco.com/Images/Mobile%20TV%20guidelines_tcm3-13804.pdf

Seyler, A. J., & Budrikis, Z. L. (1964). Detail perception after scene changes in television image presentations. *IEEE Transactions on Information Theory, 11*, 31-42.

Sling Media. (2006). *SlingPlayer mobile.* Retrieved from http://www.slingmedia. com

Södergård, C. (2003). *Mobile television—Technology and user experiences. Report on the Mobile-TV project* (Rep. No. P506). VTT Information Technology.

Song, S., Won, Y., & Song, I. (2004). Empirical study of user perception behavior for mobile streaming. In *Proceedings of the tenth ACM international conference on Multimedia* (pp. 327-330). New York: ACM Press.

Steinmetz, R. (1996). Human perception of jitter and media synchronization. *IEEE Journal on Selected Areas in Communications, 14*, 61-72.

Stockbridge, L. (2006). Mobile TV: Experience of the UK Vodafone and Sky service. Retrieved from http://www.serco.com/Images/EuroITV%20mobile%20TV% 20presentation_tcm3-13849.pdf

Taylor, A., & Harper, R. (2002). Switching on to switch off: An analysis of routine TV watching habits and their implications for electronic programme guide design. *usableiTV, 1,* 7-13.

Thompson, R. (1998). *Grammar of the shot.* Elsevier Focal Press.

Tse, T., Vegh, S., Marchionini, G., & Shneiderman, B. (1999). An exploratory study of video browsing user interface designs and research methodologies: Effectiveness in information seeking tasks. In *Proceedings of the 62nd ASIS Annual Meeting* (pp. 681-692).

Vatakis, A., & Spence, C. (in press). Evaluating the influence of frame rate on the temporal aspects of audiovisual speech perception. *Neuroscience Letters.*

Voldhaug, J. E., Johansen, S., & Perkis, A. (2005). Automatic football video highlights extraction. In *Proceedings of NORSIG-05.*

Wang, D., Speranza, F., Vincent, A., Martin, T., & Blanchfield, P. (2003). Towards optimal rate control: A study of the impact of spatial resolution, frame rate and quantization on subjective quality and bitrate. In T. Ebrahimi & T. Sikora (Eds.), *Visual communications and image processing* (pp. 198-209).

Westerink, J. H., & Roufs, J. A. (1989). Subjective image quality as a function of viewing distance, resolution, and picture size. *SMPTE Journal.*

Winkler, S., & Faller, C. (2005). Maximizing audiovisual quality at low bitrates. In *Proceedings of Workshop on Video Processing and Quality Metrics.*

Zillman, D. (1988). Mood management: Using entertainment to full advantage. In L. Donohew, H. E. Sypher, & E. T. Higgins (Eds.), *Communication, social cognition, and affect* (pp. 147-172). Hillsdale, NJ: Erlbaum.

Section III

Business and
Marketing Studies

Chapter XV

Time and Timing in Cross-Media Production:
A Case Study from Norwegian Television

Roel Puijk, Lillehammer University College, Norway

Abstract

Based on ethnographic fieldwork in two production units of the Norwegian Broadcasting Corporation (NRK) this chapter looks at the relation between the production for television and the Internet. Different organizational models can be used in cross-media production: separated production, the recycling, added value, and integrated model can be identified. The two production units under consideration are organized in different ways. This is not only caused by differences in timing of publication on television and on the Web, but also related to differences in production cycles. The Web pages have different functions for the television programs they support—depending on genre. Although officially cross-media production units, television turns out to be the main interest for the units.

Introduction

Many television corporations have become cross-media production houses, producing for radio, television, and the Internet. These activities can be organized in different ways—texts for different media can be produced by the same personnel, by different departments, or integrated in production units. Ways of organizing this production has consequences for the texts produced—sometimes the results of organizational processes are intentional, sometimes they produce unintended results. During the first season of *Idol* in Norway, for example, the production of all Internet pages concerning the program was integrated with television production. When Internet activity was taken over by a more general online entertainment unit, the journalists found out that they had "better" (i.e., more daring) stories about participants. Being separated from television production, the Internet journalists retained the same loyalties with the participants for longer (Kjus 2005).

Remediation in different media has attracted scholarly attention (e.g., Bolter & Grusin 2000), as well as interactivity through a combination of multiple platforms (Christensen, 2004; Jensen & Toscan, 1999). However, little research has been carried out on how media personnel on the ground level work: how production processes are interlinked, what ideologies are involved, how the different platforms are part of more general corporate strategies, and what part of the possibilities of the Web are used. This calls for detailed studies of production processes at the ground level. Before looking at concrete production processes, I will first consider some theoretical issues.

Production Studies

During the 1970s and 1980s a number of production studies were carried out in news organizations (for an overview see Cottle, 2000; Tuchman, 2002). Most of these were in a constructivist tradition highlighting the bureaucratic, routinized, and conservative character of news production. The production of fiction was also scrutinized, but here the focus was more often on the possibilities of creative individuals working through bureaucratic structures (e.g., Etterma & Whitney, 1982; Hirsh, 2002). Even Gitlin's (1983) critical study of prime time programming dealt with creativity. As the media quickly change there is an urgent need for what Cottle (2000) calls "a second wave of ethnographic research in production processes". As Cottle (2002) argued in the introduction to a book he edited on media organization and production:

In between the theoretical foci on marketplace dominations and play of cultural discourses, there still exists a relatively unexplored and under-theorised 'middle ground' of organisational structures and workplace practices. This comprises different organisational fields and institutional settings, and the dynamic practices and daily grind of media professionals and producers engaged in productive processes. (p. 4)

Most production studies have been carried out in American and British news organizations. Other countries and genres have attracted less attention. There seems to be cross-country variations in working arrangements. In his comparative study of newsrooms in England and Germany, Esser (1998) found different models of organization—the German journalists had a more holistic understanding of news, were more decentralized, and had less division of labor as compared to their British counterparts. When it comes to cross-media production we can also assume that there are differences in work arrangements and journalist ideologies.

Cross-Media Organizational Models

Recent studies on Internet production by existing media organizations show that different models are involved. These models involve both organizational and content aspects. Domingo (2004), who studied the professional routines in four Spanish online newsrooms, describes one model. Based on observations and interviews, he claims that being first, beating competing media in time, "*is a professional value taken to obsessive rates*" (p. 1) by online journalists. The online newsrooms in his study are rather autonomous entities with little continuity between the online and off-line newsrooms in terms of news content. The journalists hardly ever leave the newsroom because the news wire is the main, and almost only, source for most online news (Domingo, 2004, p. 9). In the case of the newspapers' Web sites, the

Figure 1. Separated model

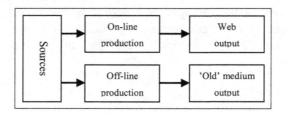

news they produce is considered provisional and replaced by "real news" produced by the traditional newsroom when the printed edition of the paper becomes available. In this situation both online and traditional newsrooms produce news, but the two newsrooms live separate lives with no continuity in content. This can be called a *separated model* (Figure 1).

This organizational model is not only in use in news production, I encountered a similar model in the NRK where the youth department produced extreme sport coverage for the Internet with hardly any relation to television production (Puijk, 2005).

Other media houses use the same material and republish it on different platforms. One development is the video reporter who, equipped with a digital video (DV) camera, microphone, and notebook is sent into the field to gather material. Back at the office the reporter is supposed to write articles, edit video, and publish it in the different media. We may call this the *recycling model* where the same material and histories are spread across different platforms (Figure 2a). In Norwegian television this model is used primarily in the regional offices.

In her study of Internet production in the Department of Culture in NRK, Dalberg (2001) shows that Internet articles are produced after the items are broadcast on radio and television (Figure 2b). She used *actor-network theory* demonstrating that although both radio and television journalists used the same software (ENPS), the

Figure 2a. Recycling model

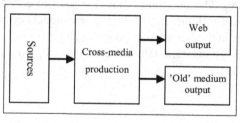

Figure 2b. Recycling model

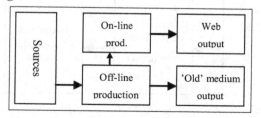

Figure 3. Added value model

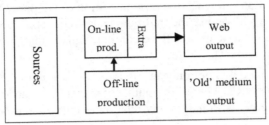

production routines in radio favored production of Internet articles, while the production of Internet articles based on television items was a cumbersome process.

Pure recycling is often looked down on as just another commercial trick, and critics often demand "added value" especially from the Internet (Christensen, 2004; Sullivan, 2004). In these cases we can call this an *added value model*, underlining that the Internet can be used to provide more information than radio and television reports on their own (Figure 3). This can be done in different forms, not only by longer background articles, but also by utilizing other Internet possibilities (linking with other sites, discussions, interactivity).

A last model is the *integrated model*. Here the Internet is established as a complementary element to television or radio broadcasts, such that the different media serve different, but complementary functions in an all-embracing concept (Figure 4). Big brother probably was the first integrated cross-media concept, involving television broadcasts, phone, and short message system (SMS) voting, Internet streaming, and chatting. The integrated model is used in entertainment programs like *Idol, Greatest Britain* (Kjus, 2005). Integrated models are also used for more educational reasons as in the Norwegian program *Pugg-and-play*—an entertaining television program for youngsters with an accompanying "homework service" on the Web (Svoen, 2005).

Figure 4. Integrated model

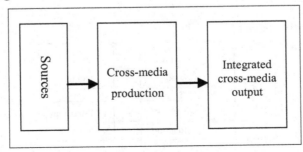

Changing Journalist Roles

Production of material for different media is not only a question of alignment between the different media. In Boczkowski's (2004) study of three newspaper Internet projects in the U.S., one of them was a community connection project, where the newspaper facilitates local groups and their use of the Internet. In this case newsroom practices were more "content management and facilitation" oriented than "reporting and editing" (pp. 141-170, 177).

Others have also noticed changing relations and journalist roles, not the least because of the possibilities of linking and interactivity on the Internet. In their study of online journalists in the Netherlands and Flanders, Deleuze and Paulussen (2002) found, in particular in Flanders, that online journalists saw their work from a "guide dog" rather than a traditional "watchdog" (p. 242) perspective. A guide dog perspective implied that a journalist's role was to help people find their way through the information on the Internet. (They concluded that:

... online journalism relates closely to [...] so-called public or civic journalism. [...] Journalistic roles such as influencing public opinion and being a spokesperson for certain groups in society are not very popular among respondents. (p. 243)

A Case Study in Cross-Media Production

In this chapter some results from an ethnographic study on cross-media production in the NRK are presented. Two production units, producing both for television, radio, and the Internet are compared. The television programs they produce are neither hard news nor fiction—one (*Puls*) is a health and lifestyle magazine program, consisting of video reports introduced and linked together by studio segments. The other (*Store Studio*) is a cultural talk show structured around a female host who invites guests from the cultural sector.

When I asked permission to carry out an anthropological study on cross-media production in NRK, the first response was not encouraging. After some repeated approaches, I was allowed to observe these two production units. The director of NRK's production departments in Oslo chose them because they were considered showcases of cross-media production. In addition, the staff of the production units was positive to my presence.

In autumn 2003 I spent 4 months in Oslo, following the daily routine and interviewing the members of the two production teams. In addition I gained access to a number of internal reports. The only thing that was kept secret or I was unable to use was information about production costs. Budgets were considered internal information; if published they might be used by NRK's competitors.

One of the main arguments in this paper is that production for Internet and radio is geared mostly towards television, but that the different genre of the programs leads to different challenges and processes. Even though none of the production units made regular news programs, timing was an important factor. The time dimension of production cycles was important for what might be called the "collective focus" in production units.

NRK's Internet site consists of more than text-based articles produced by the different production units—most self-produced television and radio programs are available on the Web, there are games, discussion fora, and many short cuts (hyperlinks) to other parts of the site (se Puijk, 2004 for a description of the structure of NRK's home pages). Here I will concentrate on the production of the articles, as they are the central focus for the relation between the Internet and television/radio production process.

The members of the production units in this case study produce their own Internet pages and have integrated Internet journalists as part of the team. In the rest of this article we shall explore how production for the Internet was related to the production for television and radio, both in organizational terms and content.

Television Production

As mentioned above, *Puls* has a typical magazine format, consisting of three video reports that are introduced and linked in short studio sequences. The program is strongly formatted: It contains one "news issue," one "lifestyle issue," and one "consumer issue," while the program ends with a sequence where an expert answers questions on health and lifestyle-related issues posed by the program's host. This pattern reflects the division of labor among the reporters—three work on news items, two on lifestyle, and two on consumer items.

In production they use single camera production technique for the reports and multi-camera for studio recordings (see Elliot, 1972). These production techniques require different manning—a journalist and photographer record on location, while, back at the office, the same journalist and a video editor cooperate to finish the sequence for broadcasting. The studio sequences are recorded by a much larger staff—program host, two camera operators, producer, script writer, sound operator, light operator, technical leader/camera controller, and graphic operator are in place to realize the multi-camera recording. The studio segments are recorded during the day, a few hours before the program is broadcast on Monday evening right after the main newscast.

The different members of the production unit are involved in synchronized, but different, production cycles. Reporters have two or three weeks to produce their items, with a deadline on Friday afternoon so the report can be incorporated in the studio recordings on Monday. After having finished their reports, or while the editor finishes editing the report, the journalist writes Internet articles accompanying the report and, if it is considered a potential news item, he/she edits a short radio item to be used on Monday in the morning newscast on the radio.

The production cycles are thus synchronized, but the members of the production unit do not share focus—the journalists focus on their reports in two/three week cycles and finish on Fridays, while other members of the unit have weekly cycles with a focus peak on Mondays. During these studio recordings—that take place in the office space—the other members of the unit work and function as a background for the recordings.

Store Studio is a studio production using multi-camera technique. Here the whole team works on the same cycles, focusing on recordings on Monday. The recording takes place in a studio with a live audience early in the evening.[2] After these are finished most members of the production unit go to the meeting room to watch the program that is aired from 10:30 to 11.00 p.m. The atmosphere is relaxed and festive at these meetings where they comment upon the broadcast in a humorous manner. SMS messages commenting on the program that some members get from friends and family are read aloud and commented upon.

Internet Activity

Puls

Puls started as a regular television program in 1999. In 2002 it was threatened by termination. In their endeavors to survive, the staff strengthened the division of labor among the journalists and became committed to extensive use of new media. In autumn 2003 the changed concept saw light, including interactivity and new Internet pages (see Puijk, 2005).

In an internal document *Puls'* values are circumscribed with the words: thoroughness, courageousness, inspiring, and closeness. Except for mentioning that the Internet pages should give an added value, *Puls'* plan of action does not say much about the Internet.

Internet activity is closely connected with the television production cycles. *Puls* has two Internet journalists who function as a desk—they receive text files from the reporters, transform them into Internet articles (dividing them into shorter articles, adding pictures, etc.), and publishing them on *Puls'* Internet pages. The articles that deal with the issues raised in the reports in the television program are published early Monday morning. Also, the radio item is published on the Monday morning news. The radio news desk accepts most items that are suggested by the *Puls* journalists, but not in an automatic fashion.

Publication in the different media is thus timed consciously—publication on the Web and the news in the morning are supposed to function as an announcement and trigger for television viewing in the evening. In particular the radio item may be picked up by other media and help to place the item firmly on the public agenda.

On October 13, 2003 one of the video reports dealt with young people's drinking habits. In the report two experts argued that it was futile to expect young people to abstain from alcohol. Instead one should teach them to limit their drinking so they obtain a "happy alcohol level" ($\approx 1‰$). There were three articles published on the Web—written by the reporter and experts on the subject, explaining how one could obtain this "happy alcohol level." Although this was a "lifestyle" issue, it was broadcast on the radio news and received much attention during the day. As a result the issue was topic of debate during a discussion program on NRK the same evening (aired 8:30-9:00 p.m.). This issue thus generated much attention and debate and it was considered in line with *Puls'* courageousness value. The journalists regarded this as a successful issue; they consider creating debate and exerting an influence as more important than abstract ratings.

In this example the Internet pages contained more background information that people could consult to find out more about the "happy alcohol level." Normally the amount of added value was restricted though—the pages not containing much more than the information shown on television.

The Internet is also used in connection with the "consumer issue." In order to stimulate interaction a consumer-topic is announced in every television program. These are topics like coldness, impotence, cold sores, and so forth where viewers can provide advice during an Internet meeting the next day. The results are the basis for Internet

Table 1. Puls' Web articles autumn 2003

Articles in connection with television broadcast (published on Monday morning)	78
Consumer issue—viewers' advice (not published on Monday)	14
Experts answer viewers' questions	28
Self–promotion	30
Following up on television issues	6
Others	14
SUM	170

articles and an item in next week's television program. As the results of the Web meeting are available on the Web, and the information is neither spectacular nor new, this material is published in Internet articles throughout the week.

Viewers can also ask questions on health-related subjects that are answered by experts. Some of these questions are answered in the television interviews, but the experts are also supposed to answer some questions in Web articles.

As can be seen in Table 1, most Web articles are the result of planned, regular activity. During their staff meetings and in interviews the members of *Puls* expressed that the Internet could and should be used more often. When a potential topic was discussed during staff meetings and considered too narrow (in terms of number of people affected) or problematic it was often added that it might be "put on the Web." The Web was also seen as a good alternative to follow up new developments in the televised topics, to provide more background information, or to voice stronger opinions than would be possible in the television program.

In reality these possibilities were seldom used—during the first weeks of the autumn season some effort was made to publish extra issues on the Web, but after a while this activity diminished and was seldom used. Reporters and Internet journalists focused on regular activity with little surplus energy to make use of these possibilities.

Store Studio

The situation for *Store Studio* is different. During the television program the female host normally interviews three guests from the cultural sector about literature, theatre, film, music, and so forth. She is assisted by a co-host who is presented as an expert on popular culture. He specializes in interviewing pop musicians, and the program normally contains short interviews. Together they have some short fun sequences during the show. The program ends with a popular band playing. The same guests and musicians are involved in the radio program, but the pace of the program is slower—the interviews more relaxed.

Although arriving a decennium later on the scene, the program reflects the changes in culture journalism in the Norwegian press—from the late 1980s and onward culture journalism changed from dealing with arts to encompassing both high and popular culture, art, and entertainment (Bech-Karlsen, 1991).

Also here, Internet activity is seen as important both for attracting viewers for the television show and in and of itself as a source of added value. In their plan of action for 2004 *Store Studio* is more specific in specifying the goals for their Internet activity. In the introduction of the plan it states: "*Internet is to build its own universe around the program and generate material that is even more daring and subjective than the radio and television program.*" This is to be obtained through what is called "tabloidization," "systematic Web publication," and *selling their topics to radio and news departments and the daily press.*"

The main goal was to fulfill the publication goals as set by the broadcaster when they commissioned the program. Another concrete goal is formulated this way: "*Store Studio* is to be experienced as necessary and important. At least two *Store Studio* topics shall be cited by other media each month."

Store Studio's Web site is organized differently as compared to *Puls*'. One Internet journalist does much the same work as those in *Puls*—finishing articles, scanning pictures and headlines, and doing the actual publishing. Several members of the production unit write for the Web, including the production assistant, production leader, researcher, host, and co-host. They are allowed and encouraged not only to write pieces in connection with guests but also to do other articles that might be of interest to *Store Studio*'s target group (young adults between 20-40 years of age).

While *Puls* publishes most of its pages the same day they broadcast the television program, in *Store Studio*'s weekly work plan Internet issues are distributed across the week—and more integrated into the preparations for the recordings on Monday. Background articles are written as part of the preparations for the program—for example, Anne's mind-map (a photo or drawing of the guest with catchwords written on it) is drawn during a staff meeting on Thursdays, when some staff members meet to discuss what themes might be discussed in the forthcoming program; the map is scanned and thereafter published on the Web. In addition background articles on the guests are published in addition to other regular pages that carry their own name ("Per's pop friends," "Strong opinions," "Anne has the floor").

The background articles on the guests have to be finished on Friday at the latest, as an e-mail is sent to subscribers of *Store Studio*'s newsletter informing them about the forthcoming program. As the plan of action states, the Web pages contain little about the television program itself—only some backstage pictures taken during the recordings are put on the Web after the television program is broadcast. The recordings are available as Web TV; both the television and the radio program are streamed with sound and picture.

As mentioned previously, one of the objectives of the Web pages is to generate debate. While most pages have rather few users, some generate much traffic. Figure 5 shows how many hits there were on *Store Studio*'s Web pages in the week from November 10-16, 2003. The pages were accessed mainly on Thursday and Friday, not on Monday when the program is broadcast. Further specification of the numbers shows that one article accounted for almost 30% of the traffic. This article, published on Thursday, was about the sale of music. The article was published under the rubric "Anne has the floor." The program host accused some of the main newspapers of having launched a "campaign to crush" artists who sell many CDs. The article is part of a more general debate about music reviewing practices and, in contrast to most NRK Web pages, it was externally linked to the newspapers in question.

Figure 5. Number of hits on Store Studio's pages, November 10-16, 2003. (Source: NRK)

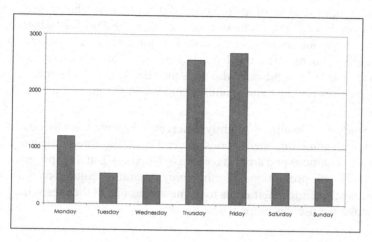

The second most read article (7%) dealt with a Nick Bromfield's film that was to be shown at the Oslo Film Festival—with no relation to the television program. It was published on Wednesday. The other pages contributed fewer hits: The backstage pictures had 5.8%, background articles for Monday's guests between 4 and 5%, while the individual pages presenting the members of the staff varied between 0.4 and 1.5%.

External and Internal Visibility

Even though NRK is financed principally through license fees, and ratings are not therefore automatically convertible to money, visibility is still one of the main objectives. The combination of media is used to achieve mutual reinforcement—television programs promote the Web pages and vice versa. But television, radio, and the Internet have a hierarchical structure as to what is most visible and what receives the most prestige. For radio and television the newscast ranks highest, while NRK's front page is most attractive for the Internet. For these there is internal competition for access. Both *Puls* and *Store Studio* try to "sell" (some of) their issues to the news and the centrally edited front page. We saw that *Puls* had a functional relation with the radio news and this allowed them to publish on average one issue each week. They also like to get their issues on the television news, but competition here (from other departments, regional offices) is fiercer and their proposals are often ignored.

During the staff meeting after the "happy alcohol percentage" program, it was commented that the host of the discussion program, who appeared on the 7 o'clock news to announce the topic of his program, did not mention the *Puls* broadcast as he had promised to do. The discussion showed how the *Puls* journalists struggled to gain acceptance for doing a good journalistic job and being equal in status with news journalists. There is also much prestige in getting an article on the front page. The Internet journalist is the one who does the "desking"—formatting the article; providing pictures and headlines; and getting credit for getting an article on the front page.

For *Store Studio* the situation is slightly different; they have few obvious hard news issues, although sometime their stories can end up in the entertainment section of the news (e.g., a famous pop artist is coming to Norway). But their program is more often part of the self-promotional "forthcoming programs" column on www.nrk.no. Still, as we saw, "selling" their items to the news was one of the means they wanted to use to make themselves more visible.

Hypertext Linking

Like most Web pages produced by media organizations, NRK's pages contain not only articles and pictures, but much space is used for links and shortcuts to other pages—different programs, such as Web radio, Web television, and archives can be reached from every page. In this way the Internet's potential for hypertext is used fully.

As show previously, discussions between media may generate many hits on the Web pages. The "campaign to crush" page was linked to the articles it commented upon and back again. In this way one could easily follow the different contributions.

Still rather few Web pages have links outside the nrk.no domain. A number of factors are involved here. First of all, time plays a role, together with quality control. Being a state-owned organization the responsible Internet journalists are restrictive when it comes to linking their pages with what might be less serious or commercial content providers. This implies that they have to check the sites they link to their articles, which of course is time consuming. Even though NRK's Internet pages contain advertisements (see Puijk, 2005), sites with obvious commercial interests are avoided. In praxis this implies that the links often refer to governmental or non-government organization (NGO) sites. Another reason for internal linking is that when people follow external links they are guided away from NRK's pages—and they might not come back. As the number of hits on the Web is one of the most important measuring rods, this becomes an impediment to sending users to other

Figure 6. The clickable body, Puls(© NRK)

domains. There are no official stated rules to guide this practice, but both quality and restriction consideration seem to be internalized by the Internet journalists.

Routine and Creativity

In his production study of a television series in the early 1970s Elliot (1972) used Burns and Stalkers concepts "mechanic" and "organic" to describe the division of labor that characterizes multi- and single-camera production. According to Elliot, multi-camera production is organized in a strictly hierarchical way with the producer at the center. The other members of the production team do standardized jobs that are coordinated centrally. In single-camera production, according to Elliot, there exists a flexible division of labor, first of all between the director and photographer, who can discuss in between the takes and subsequently adjust the recordings. This way of thinking is very much in line with how the industry thinks—multi-camera productions are standardized, rigid, and leave little room for creativity. The staff (TOM, camera operators) is defined as technical, while the single-camera productions are thought to be creative as expressed in the terms photographer and editor. Formerly, the division also was applied to different media—film versus video. Today the differences may be less—all use video—but other differences are still recognizable.

This model can be criticized for focusing on the recording period and not taking into account the whole creative process; of course, the camera operators have to be coordinated during the recordings and many multi-camera recordings, especially the weekly programs, are standardized. But also multi-camera productions can be highly creative. Most space for creativity is found during the preparation period (Puijk, 1990).

The same mechanism operates in relation to the Internet. Also here much of the day-to-day operations are routinized—Web articles often have a standardized form. But the Internet offers potentials for creativity—not the least when it comes to producing different forms. One example is the "clickable body" icon on *Puls'* Web pages.[3] By clicking on different parts of the male and female body (arm, throat, breasts, etc.) one is guided to an overview of related articles and video reports in *Puls'* archive. Compared with the usual text-based linking practice, the use of visual navigation was a creative idea put forward by the Internet journalist, using the possibilities of the medium in a new way for NRK.

Store Studio also has several Web pages (e.g., mind map, presentation of the members of staff) that represent a new way of bringing out information. Being a talk show their net pages are allowed to be more entertaining, while *Puls* pages are more serious and fit in with a "discourse of soberness."

The possibilities of the Internet are enormous, and of course both print journalists and Internet journalists are aware of these possibilities. But time and focus constraints limit the variety of forms and content produced. Routinization of production, both by incorporating assignments in the production cycle and in developing ready-made forms, facilitates production.

Production cycles in many ways determine the focus of the people involved. As we saw the multi-camera production of *Store Studio* geared the staff members towards a common focus with a climax during recordings on Monday. This resulted in a stronger team feeling as expressed in their coming together after the recordings. Several staff members produced the Internet pages, also as an accompanying contribution to the next program.

The mixed production cycles in *Puls* made it harder to integrate all staff members socially. As for Internet articles, the journalists write the pages that concern "their" issues (together with the Internet journalists). The other staff members do not contribute with articles.

Even though there are differences in production routines in *Puls* and *Store Studio* both had difficulties when attempting to realize non-routinized production: The extra articles that both mentioned not only in their plans but also during staff meetings and in interviews with me were hard to produce. Material that is not in focus as part of a routinized production cycle is much harder to accomplish. Even though the actors showed signs of good intent, the production of these extras was more erratic because the different platforms are neatly timed in relation with each other.

Conclusion

For almost half a century radio and television in NRK functioned as separate media with little contact. Today the production departments consist of cross-media teams that produce for television, radio, and the Web. Still for *Puls* and *Store Studio* television is their primary interest. Television programs are evaluated every week—Internet pages and radio programs are looked at once per season. In many ways they are television-driven production units.

Even though both production units have Internet journalists integrated in the production team, the way the Internet functions in relation to the television broadcast is quite different. The Internet pages of *Puls* in many ways follow the recycle model, although the added-value model is sought after. *Store Studio's* Web pages to a larger degree function according to the integrated model, building an entertaining background around the program and its guests.

These differences relate partly to the different genres these programs represent—one is a factual program that is based on a discourse about facts, even though not all items are critical; the other is more entertaining, mixing a focus on stardom, popular culture, and glamour with a (literature) critical one.

Timing and exclusiveness are crucial factors in the competition between media—not only in the case of news but also in case of other programs. The first time a media product is publicized its value for the receiver is the highest. This also counts for the publication of cross media content. But to be exclusive and first is not the only rule—much depends on what platform is used. For example, several television channels offer newer episodes of television fiction for money on Web TV, even though the same program may be watched for free on television the following day.

The differences between *Puls* and *Store Studio* are also connected to the timing of the media in relation with each other. Even though *Puls* Web articles are produced after the television reports, they are published beforehand and function as an announcement for the television program. The content of the television show is not known when the Internet pages of *Store Studio* are published and this prohibits the Internet pages from recycling the content of the television show. But in both cases the media are neatly timed in relation to each other.

Although the activity in both production units is geared towards television primarily, the fact that both ratings and Internet hits are measurable implies that the viewers' activity in relation to both media become critical data. Achieving not only ratings goals but also hits on their Internet pages, thus becomes important in their own right. For NRK, Internet activity in fact produces money—their Web activity is not defined as part of their public service obligation and this implies that they can use it to generate additional income. Their Internet pages thus contain advertisements, even though NRK's Internet pages are accessed free of charge. There is some uncertainty if this practice will be continued in the future.

Puls' and *Store Studio'*s journalists are not specially geared towards helping the public to obtain more information on the Internet; for this their hyperlinking is far too limited. Influencing public opinion is still an important objective; several of *Puls'* journalists can be positioned within the tradition of "critical journalism" ideology, scrutinizing public and private health organizations. On the other hand, this does not exclude that both *Puls* and *Store Studio* do not also provide their users with information—they do, but from a more "enlightenment" perspective, that is, they provide information themselves rather than referring them to other information obtainable elsewhere on the Web. In this way there is a form of quality control—the information is selected and controlled by the members of the staff (or the experts who are chosen to answer questions). In many ways one can say that this is a continuation of the organizational culture of enlightenment that formed and still forms an ideological premise of NRK (see Küng-Shankleman, 2002).

This article is based on fieldwork done in autumn 2003. At the time both the production units studied were supposed to be at the "cutting edge" of developments in NRK. Although the research was only completed a few years ago, the field is changing rapidly. New concepts, new ideologies of mediation, and new organizational forms are developing, and we may see many changes before things settle and a more permanent form is established. The head of Internet development told me in November 2005 that writing Internet articles, as *Puls'* journalists mostly do, is seen as not very appropriate. He referred to another televisions program's Web pages containing blogs as an example of what they support today. For research this implies that we have to keep looking for new developments, while at the same time not forgetting continuity, highlighting traditional elements that are also part of working with new media.

References

Bech-Karlsen, J. (1991*). Kulturjournalistikk: Avkobling eller tilkobling?* Oslo: Universitetsforlaget.

Boczkowski, P. J. (2004). *Digitizing the news. Innovation in online newspapers.* Cambrigde, MA: MIT Press.

Bolter, J. D., & Grusin, R. (2000). *Remediation. Understanding new media.* London: MIT Press.

Christensen, L. H. (2004). *Interaktivt TV .... Vent venligst...* Ålborg: Ålborg Universitetsforlag.

Cottle, S. (2000). New(s) times: Towards a "second wave" of news ethnography. *Communications, 25*(1), 19-41.

Cottle, S. (Ed.). (2002). *Media organization and production.* London: Sage.

Dalberg, V. (2001). *Representation av kontekst: flermeidal publisering på tvers av praksisfellesskap I NRK.* Unpublishes master thesis, University of Oslo, Oslo.

Deleuze, M., & Paulussen, S. (2002). Online journalism in the low countries. Basic, occupational and professional characteristics of online journalists in Flanders and the Netherlands. *European Journal of Communication, 17*(2), 237-245.

Domingo, D. (2004). *Comparing professional routines and values in online newsrooms: A reflection from a case study.* Paper presented at the IAMCR conference, Porto-Alegre, Brazil.

Elliot, P. (1972). *The making of a television series—A case study in the production of culture.* London: Constable.

Esser, F. (1998). Editorial structures and work principles in British and German newsrooms. *European Journal of Communicatinon, 13*(3), 375-405.

Ettema, J. S., & Witney, D. C. (Eds.). (1982). *Individuals in mass media organisations. Creativity and constraint.* Beverly Hills, CA: Sage.

Hirsh, E. (2002). Production of television fiction. In K. B. Jensen (Ed.), *Handbook of media and communication research: Qualitative and quantitative research methodologies.* Florence, KY: Routledge.

Jensen, J. F., & Toscan, C. (1999). *Interactive television. TV of the future or the future of TV?* Denmark: Ålborg University Press.

Kjus, Y. (2005). *Idol: Formatet, mediene og publikummet.* Paper presented at the conference Fjernsyn i digitale omgivelser, Lillehammer, Norway.

Küng-Shankleman, L. (2002). Organisational culture inside the BBC and CNN. In S. Cottle (Ed.), *Media organization and production.* London: Sage.

Puijk, R. (1990). *Virkeligheter i NRK.* Lillehammer: Private Publication.

Puijk, R. (2004). Television sports on the Web: The case of Norwegian public service television. *Media, Culture and Society, 26*(6), 901-910.

Puijk, R. (2005). The use of Internet in television—A case study from Norwegian public service broadcasting. In J. F. Jensen (Ed.), *Proceedings: 3rd European Conference on Interactive Television: User Centred ITV Systems, Programmes and Applications.* Denmark: Ålborg University.

Sullivan, J. L. (2004). *In the rewards and perils of "studying up." Practical strategies for qualitative research on media organisations.* Paper presented at the IAMCR conference, Porto Alegre, Brazil.

Svoen, B. (2005). Children's and youth television and the convergence of media: A look at the interactive cross-media case "puggandplay." In J. F. Jensen (Ed.), *Proceedings: 3rd European Conference on Interactive Television: User Centred ITV Systems, Programmes and Applications*. Denmark: Ålborg University.

Tuchman, G. (2002). The production of news. In K. B. Jensen (Ed.), *Handbook of media and communication research: Qualitative and quantitative research methodologies*. Florence, KY: Routledge.

Endnotes

[1]   The chapter is a result of the Television in a Digital Environment project (see: TiDE.hil.no), financed by The Research Council of Norway.

[2]   The recording is divided into two—one half hour recordings for television broadcast 10:30-11:00 p.m., one half hour for radio, broadcast immediately after (11:00-11:30 p.m.). The radio recordings are also videotaped and broadcast on television during Christmas and Easter.

[3]   http://www6.nrk.no/programmer/puls/mann/kroppen_mann.html (Retrieved June 18, 2006)

Chapter XVI

From 'Flow' to 'Database':
A Comparative Study of the Uses of Traditional and Internet Television in Estonia

Ravio Suni, University of Tartu, Estonia

Abstract

This chapter compares several basic statistical indicators of broadcast (traditional) television viewing and Internet protocol television (IPTV) use in Estonia and show how the structural difference between the two types of television results in different consumption models. The main conclusion is that the structure of the content to a large extent determines the uses of media. Flow-type media (broadcast television) appears to support routine and unconscious media use, while the use of database-like media (Internet television) could be characterised as being more purposeful and conscious. The study opens new perspectives on audience research and might be inspiring for further research and analyses focussing in detail on the use of IPTV.

Introduction

Although most of us watch television every day, it is hard to define it unequivocally. For TV professionals, television is usually a text that is constructed of moving pictures and sound, and that should somehow attract the viewers. For viewers, television is above all a moving picture on a TV set, situated in the living room; it entertains, passes the time, or provides interesting and useful information. Communication and cultural scientists usually sum up these two understandings and interpret television as a form of culture that constantly picks up some pieces of culture, processes them in its own way, and gives them back to the culture in a slightly different form (see Morley, 1992; Williams, 1992).

Despite the fact that different groups have interpreted television differently, the general understanding of TV was quite homogenous in society up to the 1980s. The consumption context was quite similar for most individuals—television was usually watched at home (in the living room), after school or work, often together with other family members (see Ang, 1991, 1995; Lull, 1991). The choice was limited to one or two TV channels.

All this determined much of the content that TV producers offered—mostly the programmes were of general interest to a general audience, news was aired at the best prime-time slots, while adult programming was aired late at night when children had been put to bed.

Such a homogenous understanding of television began to disappear in the mid-1980s, when the TV industry entered an age of uncertainty (Ellis, 2000). The liberalisation of media markets and the development of new communication technologies pushed media companies to exploit new audiences.

The targeting of new television audiences achieved a new level in the era of Internet and mobile phones. Television dematerialised from the *apparatus televisio* and appeared everywhere—on Internet-connected PCs, on laptops situated in a wireless Internet area, or on a mobile phone.

Figure 1. Bordewijk and Kaam's matrix for the four communication patterns. Source: Bordewijk and Van Kaam (1986)

	Information produced by a central provider	Information produced by the consumer
Distribution controlled by a central provider	**TRANSMISSION**	**REGISTRATION**
Distribution controlled by the consumer	**CONSULTATION**	**CONVERSATION**

It is hard to identify the first Internet TV experimentation, as the software development took place simultaneously in different parts of the world. In 1995, a U.S. computer magazine *Computer Chronicles* was accessible over the Internet (Greenberg & Johnson, 2004). In October of the same year, when the Pope visited the U.S., Xing Technology Corporation and Catholic Internet made an experimental live broadcast of this event over the Internet.[2] In February 1996 British radio station Talk 101 made music videos available over the Internet.[3] In 1997, CNN, Fox News, BBC, and many other TV stations had news videos uploaded onto the Internet.[4]

Estonia was among the pioneers of IPTV. As early as 1998 it was possible to watch some public service Estonian television programmes live on the Web (for a detailed history of IPTV in Estonia see Suni, 2005).

Today there are two IPTV providers in Estonia, itv.ee and tv.ee. The first of them offers mainly Estonian television programmes, the other programmes from the private commercial television channels Kanal 2 and TV3. The penetration rate of IPTV is relatively high—22% of the Estonian population have used the Internet to watch television programmes.[5]

Use Determined by Structure

The development of new media has primarily forced us to reconsider the cultural meaning of television. The starting point here could be the paradigmatic changes in the television communication model.

To understand these changes it is useful to take a look at the matrix of communication patterns provided by Bordewijk and Van Kaam in 1986 (see Figure 1). Their typology is based on two central aspects of all information traffic: (1) the question of who owns and provides the information, and (2) who controls its distribution in terms of timing and subject matter (see also Jensen, 1999; McQuail, 2000).

According to Bordewijk and Van Kaam (1986) broadcast television is based on the transmission model, that is, one-way communication, where the significant consumer activity is pure reception. The key concept of the distribution of content in broadcast television is a "flow"—a continuous succession of programmes (Williams, 1992).

However, IPTV is based on the consultation model of communication. It is a type of on-demand service, where the content is produced and owned by a television company, but the consumer determines the context of reception (i.e., what, when, and where to watch). The only limit is usually access to content (the availability of programmes, Internet connection, etc). The content of IPTV is usually distributed as a "database," where the programmes are separated from each other and can be used independently.

Thus we could define IPTV as a limited form of interactive television (iTV). It is a prototype of a video-on-demand system that relies on participation and requires the user to be more active than does broadcast television.

To understand the possible differences between broadcast and IPTV use we must first of all consider their structural differences. Structural factors are universal agents that have an impact on the use of media. They depend on the nature of the medium. Institutional factors could also be called unavoidable conditions, and they are usually the same for everyone.

Structure of content. The content and viewing experience of broadcast television is very strongly connected to real time. Television is a continuous flow of programmes, programmed and scheduled by a television company. However, the IPTV is programmed and structured as a database where the user can decide what to use and when to use it. The only limitation is the access to or the availability of the programmes.

Time of use. Broadcast television viewing is a synchronous activity. The distributor usually decides when one can watch. Any interruption in viewing automatically means the loss of information. However, IPTV use is asynchronous. The user decides when to use it. The consumption of the content could be stopped and/or repeated an unlimited number of times.

Freedom of choice. With broadcast television, people are pushed to watch the "flow" of programmes provided by the central distributor. The viewing is polysemic (e.g., TV as background, companion, etc.). Interest in certain TV programmes is usually formed operatively during the general viewing process. In contrast, IPTV provides the opportunity to pull programmes and the interest in a particular content is usually there before the viewing process. People watch programmes rather than channels.

Environment of use. In spite of the increase in the number of multiple-TV homes, television has still maintained the status of a collectively consumed medium, viewed "passively" and from a distance while sitting in a comfortable chair (Stewart, 1999, p. 236). Even if there are many TV sets at home, the main TV set is usually situated in the living room. The choice of channels usually depends on the hierarchy of the viewers and negotiations between family members. In contrast, the IPTV demands a high level of engagement and interaction with the content, and is used individually. Although Internet use still takes place mainly in the public space (at work places, schools, universities, libraries, etc), there is a trend towards the rapid domestication of this media (Cummings & Kraut, 2002). In the home, Internet-connected computers are usually situated in a study.

Participation activity of audience. Despite the recent move towards more interactive programme formats, television viewing still remains in general a passive activity. The viewing does not require any specific knowledge on the part of the viewer. The activity is pure reception—receiving messages and interpreting them. In contrast, the use of the Internet and IPTV embodies quite a lot of activity in terms of navigation of the system. It demands computer and Internet literacy from the user. The user has to understand the logic of the content, to know the principles of use, and to make lots of decisions.

This was a quick and generalised theoretical comparison of transmission and consultation television communication systems. These differences probably also create the differences in the use of flow- and database-like television. The aim of the research is to describe and analyse these differences in the case of Estonia. There are several characteristics of both broadcast and IPTV that I wish to pay attention to in particular: (1) daily reach, (2) duration of use, (3) round-the-clock division, (4) individuality of use, and (5) place of use.

In addition to the general description of data I will present the more common reasons for IPTV use.

Methodology

To study comparatively the use of broadcast and IPTV, I have used two sets of data. As a base I will combine the results of my own Web-based survey conducted in June 2004. Alongside this, I will use television audience research from the market survey company TNS EMOR and the study "Me. World. Media" (MEEMA) conducted in 2002 by the Department of Communication and Journalism of Tartu University.

IPTV

The need to access the maximum possible number of IPTV users determined the selection of the method. According to a study conducted by the University of Tartu's Department of Journalism and Communication in late 2005, 22% of Estonian residents have watched TV shows via the Internet at least a couple times per month. It would have been very complicated and costly to form a representative sample considering this small number of users.

Instead of that I used the combined method of snowball sampling techniques and Web-based survey. The idea behind snowball sampling is to look for networks of identified respondents to find potential contacts (Thomson, 1997; Vogt, 1999). In my work I used the database of registered users of itv.ee to find people who have had contact with IPTV. To get in touch with them, I used a Web-based questionnaire.

The Web-based questionnaire is a method with both strengths and weaknesses. The strengths are: (1) low cost of research (Dillman, 2000; Shmidt, 1997; Shannon, Johnson, Searcy & Lott 2001; Umbach, 2004; Watt, 1999); (2) savings of time in collection and initial processing of data (Berge & Collins, 1996; Schmidt, 1997; Umbach, 2004; Zhang, 1999); (3) smaller number of coding errors (Umbach, 2004;

Zhang, 1999); and (4) possibility of hitting groups that are relatively marginal and to which access would be limited were other methods used (Coomber, 1997; Umbach, 2004; Zhang, 1999).

The weaknesses of Web-based questionnaires are: (1) coverage error—mismatch between the target population and the frame population (Couper, 2000; Umbach, 2004); a greater chance of responses from those more active in society; (2) sampling error—the conventional wisdom is that to obtain an objective result the whole population should have equal opportunity to be included in the sampling (Dillman & Bowker, 2001; Umbach, 2004). A Web-based survey would exclude those who lack Internet access; (3) non-response bias—the bias that is introduced when respondents to a survey are different from those who did not respond in terms of demographics or attitudes (Sax, Gilmartin, & Bryant, 2003; Umbach, 2004; Zhang, 1999).

The shortcomings of the Web-based survey as a method, as detailed previously, force us to devote continuous consideration to what it is possible to analyse on the basis of data collected in the Web survey as well as what it is not possible to analyse. It is not possible to make generalisations about the entire Estonian population on the basis of the analysis obtained. It is also not possible to make generalisations about all Internet users. Instead, it shows certain trends among a large number of IPTV users.

The Web-based survey was conducted in June 2004. The questionnaire consisted of 40 questions with 138 characteristics concerning different characteristics of IPTV use (the time, place, motives, habits, etc.). A personal e-mail was sent to approximately 1,500 registered users of itv.ee asking them to participate in the survey. The e-mail also included a request to fill in the questionnaire within a week on the Web page ‖http://uuring.etv.ee.

As a result of the posting, I received 103 e-mails that advised that the e-mail address in itv.ee database was incorrect and my request could not be delivered. During the week I received 673 forms, 631 of which had more than 50% of the questions filled in and could be used for the study (i.e., 42% from the potential sample). The response rate could have been a slightly higher; however, 631 forms are enough to draw the main conclusions.

Broadcast Television

The data to describe the trends of broadcast television were mainly obtained from TNS EMOR, which regularly collects data on television viewing preferences. Since September 2002 the research company has used *peoplemetres* as the main method of gathering information—195 households (ca 460 people) are included in the sample. The sample is chosen randomly and is intended to be representative of the whole of Estonia.

In addition to TNS EMOR, I use some comparative data from the study of the University of Tartu. From December 2002 to January 2003 the Department of Journalism and Communication conducted a representative survey of Estonia (MEEMA). The questionnaire was filled in by 1470 respondents and included 799 characteristics about different aspects of personal life, value orientation, habits, and so forth. There were also some questions about media habits and specifically about television viewing.

Results

I have divided this section of the chapter into three parts. First I will give a comparative description of general uses of IPTV and broadcast television (the results are presented in Table 2). In the second part I introduce the main reasons for IPTV use. Finally, I will discuss the possible impact of IPTV on broadcast television viewing.

Table 2. The comparative study of broadcast and IPTV use. Source: MEEMA, 2002; EMOR, 2005; & Suni (2005)

	Broadcast TV (among TV-set owners, %)	Internet TV (among IPTV users, %)
Daily reach	90	9
Duration of use 　Less than 30 min 　More than 2 hrs	3 52	43 3
Round-the-clock use* 　Morning (06-10) 　Day　　(10-18) 　Evening (18-24) 　Night　　(24-06)	24 53 71 18	2 8 43 11
Individual vs. Collective** 　Individual 　Collective	41 59	73 27

* *The reach for broadcast television, and of the response "use often" for IPTV*

** *The share for broadcast television, and the frequency (share) of responses "use often" for IPTV*

Television Viewing vs. IPTV Use

Daily reach. According to MEEMA, 90% of TV-set owners in Estonia watch television every day. For many people TV viewing seems to be a kind of physiological need, it is done routinely day in and day out.

At the same time the daily reach of IPTV is only 9% (the weekly reach is 48% and monthly reach 82%). Eighteen percent of IPTV users watch IPTV less than once a month. One of the reasons for the low daily reach could be the newness of IPTV. However, a more relevant reason probably lies behind the structural factors of IPTV.

When using IPTV people have to choose every single programme they want to watch. The continuous process of selection can be quite exhausting, because users have to be constantly active. In general, it means that IPTV is better suited to satisfying very specific needs and interests. This viewpoint becomes even more evident when we compare the duration of broadcast television viewing and IPTV use.

Duration of use. According to MEEMA, 3% of people watched television for less than 15 min, and 52% of people watched television for more than 2 hours at one time. The flow structure guarantees that there is always something running on TV. The dominant criteria for programme choice is related rather to the question "what to watch on TV?" rather than to the question "to watch or not to watch a programme?" The latter question is more characteristic of IPTV.

IPTV use is a much shorter experience. Forty-three percent of users watch IPTV for less than 30 minutes at one time, and only 3% of IPTV users watch TV programmes from the Internet for more than 2 hours at one time.

Round-the-clock division. According to TNS EMOR, in September 2005 the reach of morning viewing was 24%, day viewing 53%, evening viewing 71%, and night viewing 18%. Comparing this data with responses to the question "How often do you use IPTV in the morning, during the day, in evening, and at night?" we see some similarities between the round-the-clock use of broadcast and IPTV.

In general, daily IPTV use mirrors broadcast television viewing. Just as most people watch television in the evening, they prefer to use IPTV in the evening as well. However, compared to broadcast television viewing the users of IPTV tend to watch fewer programmes during the day and more at night.

Here we have to consider the fact that TV viewing is based on the statistics of peoplemeters but the IPTV use day model is based on the opinions of interviewees. This could lead us to suspect that some IPTV use that was claimed to have taken place at night might have actually taken place in the late evening. But the trend seems to be that today the use of IPTV is spread more equally around the clock than broadcast television viewing.

Individual versus collective use. Broadcast television viewing has usually been a collective activity. However, this trend is gradually changing. According to TNS EMOR peoplemeter survey, 41% of TV-viewing time was spent alone in front of TV set in September 2005.[6] The share of collective viewing was 59%.

As with computer use in general, IPTV use is more an individual activity. In the study I distinguished two groups of users: (1) those who often use IPTV alone (here I have excluded those respondents who often use IPTV both alone and together with others), and (2) those who often use IPTV together with family members and friends.

The share of those who often use IPTV alone is 73%, and of those who often use IPTV together with others 27%. The relatively high rate of collective viewing is a little surprising. However, it confirms that television viewing (either via broadcast TV or via IPTV) is often a social activity.

Place of use. As there are more and more places where TV can be watched, the importance of out-of-home viewing has gradually increased. Unfortunately, there are no relevant studies about out-of-home television viewing in Estonia. Therefore, I rely on data from other countries.

In 2005 the U.S. trial of the Portable People Meter (PPM) in Houston found that 13% of total TV viewing occurs outside the home.[7] Thus we can assume that somewhere between 85-90% of television viewing takes place at home. Broadcast television is still quite a domestic activity, though new technologies pull viewers more and more to watch TV in places other than the home.

Whether IPTV viewing is more of a domestic or an out-of-home activity should depend significantly on the domestic use of computers and the Internet. Despite the fact that Estonians, in general, mainly use the Internet outside of the home (68% use the Internet at work or school, 44% at home[8]), the use of IPTV is very dominantly domestic. Ninety percent of "use often" users use IPTV more often at home than outside the home, and only 10% prefer to use IPTV more often outside home (at work or at school, at a friend's place or in libraries).

The figures from both broadcast television and IPTV show that television viewing in general is a highly private activity and belongs to the private sphere.

Reasons for the Use of IPTV

The daily reach and duration of IPTV use indicates that IPTV is usually used to satisfy specific interests and needs. This assumption can be confirmed when we look at the different reasons for IPTV use (see Table 3).

People use IPTV most often to watch archive programmes, that is, TV programmes that are no longer aired. The second and third most popular reasons are the search for excitement and the desire to watch missed programmes that run on TV. The latter reason is slightly stronger than the first (a missed programme is often a reason for 33%, and the search for excitement is often a reason for 26% of users).

Sixty-six percent of respondents use IPTV to watch programmes that they have already seen on TV (for example the previous day). However, this is a rather rare practice. Only 7% of respondents said they often watch the same programmes on IPTV that they have already seen on broadcast television.

Thus, we can roughly distinguish three dominant psychological factors that lie behind the most frequent reasons for IPTV use. These are nostalgia, consistency, and curiosity.

The term *nostalgia* here refers to the happy memories of pleasant media experiences. The online archive of television programmes enables people to watch TV shows that were their favourites some years or decades ago and re-experience the same emotions they once had. Nostalgia could stimulate the use of IPTV directly and indirectly. In the first case people satisfy their own needs. In the latter case, they wish to share their pleasant "media memories" with others (mainly with children), suggesting that they watch the same programmes they once watched and liked.

Consistency refers to an individual's personal need or wish "to maintain a complex and precarious connectedness and coherence among his or her inner experiences"

Table 3. Reasons for using IPTV (% of responses, N=631)

Reason	Often, sometimes	Never
To watch archive programmes, that is, programmes that are not shown on TV anymore	90	10
To find something exciting there	87	13
To watch programmes that I missed on TV	86	14
While surfing I accidentally find myself using a page where there are videos and TV programmes	82	18
To pass time, because I have nothing to do	66	34
To satisfy curiosity aroused by a link that I found from the Internet or that was sent to me by friends	66	34
To watch programmes I saw on TV some time ago (yesterday)	62	38
To find out about the possibilities IPTV offers	59	41
To watch favourite programmes that I regularly watch on the Internet	47	53
To keep informed about life in Estonia when I am abroad	38	62
I can combine my computer work with TV viewing	35	65
Because my work or school work demands it	29	71

(McGuire, 1974, p. 174). According to the consistency theory, any new information might well upset the deliberate balance one has achieved. In this case, IPTV functions as a stabiliser of a personal communication system. It gives people another opportunity to obtain the information he or she missed last time.

Curiosity can be defined here as an interest in inquiry. This is a part of personal communication in which the object of discussion is a television show (actors, context, etc). Curiosity as a phenomenon is closely related to the consistency aspect of IPTV use. However, instead of a personal (existential) need to be informed, curiosity refers to an extra value that can be obtained from accidental information.

IPTV's Impact on Broadcast Television Viewing

The increasing availability of programmes from the Internet has had some impact on the habits of broadcast television viewing. Twenty-six percent of IPTV users recognise that they have started to watch less broadcast television because of IPTV. Especially significant are changes among younger people. Thirty-five percent of IPTV users aged 30 or under say they have begun to watch less broadcast television.

Worth highlighting is the correlation between the amount of broadcast television viewed and the feeling that there was a decrease in that television viewing because of IPTV. Those who count themselves as heavy television viewers are less sceptical about IPTV's impact. Four percent of heavy viewers are sure that they watch less TV today than before the IPTV era. Among light viewers the same figure was 15%.

It seems that heavy viewers are a quite stable group. For them, IPTV is more like an experiment that could be used sometimes, but broadcast TV will remain their main medium.

Five percent of respondents are sure that IPTV is addictive, and 26% think it might be addictive. The trend seems to be that light-TV viewers believe more in possible addiction than do heavy viewers. Women believe more than other demographic groups that IPTV could be addictive.

Conclusion

Many are sceptical about calling the Internet a mass medium, for it offers rather different communication experiences (McQuail, 2000; Morris & Ogan, 1996). But what are we to think about video clips that can be accessed through the Internet and which therefore should theoretically offer a relatively similar communication experience to most of the users?

The discussion would become easier if we were to reject clear criteria that have been the basis of Morris and Ogan's (1996) approach. Instead, the development of mass media should be observed as a continuous process.

One of the first to describe the contents of this process was McLuhan (1964) and Bolter and Grusin (1999) gave it the title of *remediation*—presence of one medium in another. This theory's strength comes from the fact that it touches the differences and similarities of old and new media at the same time.

By Bolter and Grusin's (1999) definition it can be said that IPTV is one of the re-mediated forms of broadcast television. If the medium created by the remediation features standardised contents, and if it is accessible and used by a mass audience, it can be defined also as a mass medium.

The extent to which new communication media may replace existing media has been a concern of media researchers for some time (Althaus & Tewksbury, 2000). History has proved that the introduction of new media technologies usually increases audience fragmentation (Dizard, 1997; Webster, 1989). However, that is not the only result.

This chapter focused on the use of two different types of television—flow-like television, which is characterised by the transmission communication model, and database-like television, which represents the consultation communication model. The aim was to find out whether different communication models also cause differences in general consumption models.

The comparative study of the uses of broadcast and IPTV open new perspectives in audience research. First of all, we can perceive the paradigmatic change in the perception of television viewing in general.

When we consider the flow-like distribution model, we are used to talking about "watching television." Although some of the programmes are selected from the TV schedule consciously and intentionally, most of the television viewing is just continuous and accidental. However, from database-like IPTV people tend to watch specific segments, or programmes. The selection of IPTV programmes is usually rational, calculated, and stimulated by specific situative needs (nostalgia, excitement, the need to be in touch with "important" issues, etc). Instead of "watching" we rather "use" or "check" the television.

As we saw from the analysis, the structure of the content to a large extent determines the uses of media. Flow-type media appear to support routine and unconscious media use, while the use of database-like media could be characterised as more purposeful and conscious. People watch IPTV with a particular reason in mind. Purposeful and focussed (active) media use is usually an experience of short duration at any one time.

Only 9% of IPTV users watch TV programmes over the Internet every day. About 43% of IPTV users spend less than 30 consecutive minutes on IPTV, while 52% of viewers watch flow-like television for more than 2 hours.

Whereas, in the early days, television viewing was considered mainly a collective activity, today more and more people prefer to watch television alone. It has been assumed that computers and the Internet tend to increase individuality even more. Through its capabilities, the Internet draws people away from family and friends (Nie, Hillygus, & Erbring, 2002).

Interestingly, the results of the survey show that the Internet and IPTV do not have to be an individual activity at all. Twenty-seven percent of IPTV users indicated that they watch IPTV more often with family members and friends than alone.

This demonstrates the importance of media use as a social activity in general. Despite the fact that the whole context of Internet use is very individualistic, people still find a way to use that individualistic medium to maintain and develop their social network.

In spite of the structural differences between broadcast and IPTV and of the ability to watch IPTV whenever and wherever a computer is connected to the Internet, television viewing or "checking" remains a domestic activity. Even when people in Estonia use the Internet more often in the workplace or at school, almost 90% of IPTV users use IPTV more often at home than outside the home.

To conclude the comparative study of television as flow and television as database, we can distinguish two types of general practices.

On the one hand, there are flexible practices (frequency and duration of media use, individualisation, etc.), which depend more on the structure of the medium. On the other hand, there are some more rigid practices (domestic use, sociality, round-the-clock division, etc.), which similarly characterise both broadcast television and IPTV.

Limitations

While discussing the differences between broadcast and IPTV we must also consider some limitations. First, IPTV as a term is much more ambiguous than implied by this paper (see Noll, 2004). In this work I discussed IPTV as one form of IPTV where a wide audience has access to one environment (a video portal[9]) and to the programmes stored there. The video portals active in Estonia do not offer live stream and therefore IPTV is mostly a television archive for local users where shows are stored after they have been broadcast on television.

We also have to consider that IPTV is in a phase of active development and many assessments and statements made today may prove inaccurate in a couple of years, requiring reworking or adjusting.

References

Althaus, S. L., & Tewksbury, D. (2000). Patterns of Internet and traditional news media use in a networked community. *Political Communication, 17,* 21-45.

Ang, I. (1991). *Desperately seeking the audience.* London: Routledge.

Ang, I. (1996). *Living room wars: Rethinking media audiences for a postmodern world.* London: Routledge.

Berge, Z. L., & Collins, M. P. (1996). IPCT journal: Readership survey. *Journal of American Society for Information Science, 47*(9), 701-710.

Bolter, J. D., & Grusin, R. (1999). *Remediation. Understanding new media.* Cambridge, MA: The MIT Press.

Bordewijk, J. L., & Van Kaam, B. (1986). Towards a new classification of teleinformation services. *Inter Media, 14*(1), 16-21.

Coomber, R. (1997). Using the Internet for survey research. *Sociological Reserach Online, 2*(2). Retrieved May 30, 2006, from http://www.socresonline.org.uk/2/2/2.html

Couper, M. P. (2000). Web surveys. A review of issues and approaches. *Public Opinion Quarterly, 64*(4), 464-494.

Cummings, J. N., & Kraut, R. (2002). Domesticating computers and the Internet. *Information Society, 18*(3), 221-233.

Dillman, D. A. (2000). *Mail and Internet surveys. The tailor design method* (2nd ed.). New York: Wiley.

Dillman, D. A., & Bowker, D. (2001). The Web questionnaire challenge to survey methodologists. In U.-D. Reips & M. Bosnjak (Eds.), *Dimensions of Internet science.* Lengerich, Germany: Pabst Science.

Dizard, W. (1997). *Old media/new media: Mass communications in the information age* (2nd ed.). White Plains, New York: Longman.

Ellis, J. (2000). *Seeing things. Television in the age of uncertainty.* London: I. B. Tauris Publishers.

Greenberg, G., & Johnson, C. (2004). *Digital times at ridgemont high. Exploring the instructional value of new media delivery systems.* Paper presented at the 20th Annual conference on distance teaching and learning. Retrieved May 31, 2006, from http://www.uwex.edu/disted/conference/ Resource_library/proceedings/04_1325.pdf

Jensen, F. J. (1999). The concept of "interactivity" in "interactive television" and "interactive media." In J. F. Jensen & C. Toscan (Eds.), *Interactive television. TV of the future or the future of TV.* Denmark: Aalborg University Press.

Lull, J. (1991). *Inside family viewing.* London: Routledge.

McGuire, W. (1974). Psychological motives and communication gratification. In J. G. Blumler & E. Katz (Eds.), *The use of mass communication. Current perspectives on gratifications research*. Beverly Hills, London: Sage.

McLuhan, M. (1964). *Understanding media. The extensions of man*. New York: McGraw-Hill.

McQuail, D. (2000). *McQuail's mass communication theory* (4th ed.). London: Sage.

Morley, D. (1992). *Television, audiences and cultural studies*. London: Routledge.

Morris, M., & Ogan, C. (1996). The Internet as mass medium. *Journal of Communication, 46*(1), 39-50.

Nie, N., Hillygus, S., & Erbring, L. (2002). Internet use, interpersonal relations and sociability: Findings from a detailed time diary study. In B. Wellman (Ed.), *The Internet in Everyday Life* (pp. 215-243). London: Blackwell Publishers.

Noll, A. M. (2004). Internet television. Definitions and prospects. In E. Noam, J. Groebet, & D. Gerberg (Eds.), *Internet television*. London: Lawrence Erlbaum Associates.

Sax, L. J., Gilmartin, S. K., & Bryant, A. N. (2003). Assessing response rates and nonresponse bias in Web and paper surveys. *Research in Higher Education, 44*(4), 409-431.

Schmidt, W. C. (1997). World wide Web survey research. Benefits, potential problems and solutions. *Behavioral Research Methods, Instruments and Computers, 29*(2), 274-279.

Shannon, D. M., Johnson, T. E., Searcy, S., & Lott, A. (2001). Using electronic surveys. Advice from professionals. *Practical Assessment, Research and Evaluation*, 8(1). Retrieved May 30, 2006, from http://pareonline.net/getvn.asp?v=8&n=1

Stewart, J. (1999). Interactive television at home: Television meets the Internet. In J. F. Jensen & C. Toscan (Eds.), *Interactive television. TV of the future or the future of TV*. Denmark: Aalborg University Press

Suni, R. (2005). The impact of contextual factors on habits of internet television use. In J. F. Jensen (Ed.), *Proceedings of EuroITV2005: User-centered ITV systems, programmes and applications*.

Thomson, S. (1997). Adaptive sampling in behavioural surveys. *NIDA Research Monograph,* 296-319.

Umbach, P. D. (2004). Web surveys. Best practices. *New Directions for Institutional Research, 121,* 23-38.

Watt, J. H. (1999). Internet systems for evaluation reserach. In G. Gay & T. L. Bennington (Eds.), *Information technologies in evaluation. Social, moral, epistemological, and practical implications.* San Francisco: Jossey-Bass.

Webster, J. G. (1989). Television audience behavior: Patterns of exposure in the new media environment. In J. Salvaggio & J. Bryant (Eds.), *Media use in the information age: Emerging patterns of adoption and consumer use.* Hillsdale, NJ: Erlbaum.

Williams, R. (1992). *Television. Technology and cultural form.* London: Wesleyan University Press.

Vogt, W. P. (1999). *Dictionary of statistics and methodology: A nontechnical guide for the social sciences.* London: Sage.

Zhang, Y. (1999).Using the Internet for survey research. A case study. *Journal of American Society for Information Science, 51*(1), 57-68.

Endnotes

[1]   Estonia is a country in Northeastern Europe. It has land borders with Latvia and Russia, and is separated from Finland in the north by the Gulf of Finland. The total area of the country is 45,255 km2 and the population of Estonia is 1,344,000 (2006). Estonia has been a member state of the European Union since May 1, 2004. Estonia is among the leading countries in Europe by internet penetration – according to the TNS Emor almost 60% of Estonian population are online.

[2]   Xing Technology Corp. Unveils revolutionary stream propagation product; mass network distribution technology added to StreamWorks product line. *Business Wire.* October 17, 1995.

[3]   Video 'netcasting' is making strides online. *Billboard.* March 2, 1996, 108(9).

[4]   Webcasting and convergence: Policy Implications. Retrived May 30, 2006 http://www.oecd.org/dataoecd/12/13/2091391.pdf.

[5]   From survey conducted by the Department of Journalism and Communications of the University of Tartu (November 2005)

[6]   The data describes the viewing habits of households with more than one person

7   *from news of Media in Canada* "U.S. PPM trial includes podcasts and finds 13% of TV is O-O-H; Canadian PPM finds 12% more tuning". Retrieved October 6, 2005, from http://www.mediaincanada.com/articles/mic/20050906/ppm. html

8   Data provided by Kristina Randver from TNS EMOR (November 17, 2005).

9   The term "video portal" is widely used in Estonia instead of the term "Internet television".

Chapter XVII

Belgian Advertisers' Perceptions of Interactive Digital TV as a Marketing Communication Tool

Verolien Cauberghe,
Faculty of Applied Economics, University of Antwerp, Belgium

Patrick De Pelsmacker,
Faculty of Applied Economics, University of Antwerp, Belgium

Abstract

This chapter investigates the knowledge, perceptions, and intentions of advertising professionals in Belgium toward the introduction and use of interactive digital television (IDTV) as a marketing communication tool. In total, 320 marketing professionals cooperated in a Web survey that was posted just before the commercial launch of IDTV in Belgium. The results show that their knowledge concerning the possibilities of advertising on IDTV is very limited, but their intentions to use IDTV in the future are relatively promising. Among the major perceived advantages of the medium are the possibility to provide more product information, two-way communication with the consumer, and the ability to target the audience more specifically. The major perceived disadvantages of the medium are the general lack experience of using it, the low adoption of IDTV by the end user, and the high cost for the advertiser.

Introduction

The convergence of three industries (Negroponte, 1995), namely content (entertainment, publishing, advertising, ...), telecommunications, and computing, made possible by digitalization, goes hand in hand with a divergence of new media devices. IDTV is the most visible result of this convergence offering new opportunities, but also implicating new threats for marketers and advertisers because the existing TV business model becomes unstable, certainly for broadcasters. The latter face the threat that new entrants deliver content directly to the telecommunication provider bypassing the packaging function of the broadcasters (Pagani, 2000). To face the major business issues initiated by new digital technologies like video on demand (VoD) and the personal video recorder (PVR), Wirtz and Schwartz (2001) recommend that broadcasters should cooperate and embrace these new developments to hold stance. As a consequence of these technological changes, broadcasters are investigating strategies to ensure future revenue streams. *"The economics of convergence will require people to pay for what they get"* (Dennis, 2002, p. 10), through subscription fees for services and channels or pay per view, although advertising revenues can lower the barrier to adopt IDTV by the end user. The adoption of the new associated applications of IDTV, for example, VoD, PVR, and the electronic program guide (EPG), by the end user grows slowly, but consistently. In 2004 14% of European households owned an IDTV (IPSOS, 2004). In several countries this adoption rate is higher, for example, in the UK 63% of the households make use of the interactive services provided by TV broadcasters and telecom operators (Ofcom, 2005). This growth trend is making IDTV attractive for advertisers in terms of reach (Ducoffe, Sandler, & Secunda, 1996). Leckenby (2003) emphasizes that in addition to the *adoption* of the technology by the end user, the success of a new medium—IDTV in this case—also depends on the willingness of advertisers to invest in the new medium. This study was carried out just before the launch of IDTV in Belgium (May 2005). It is therefore exploratory and descriptive in nature and no formal hypotheses are advanced. In this study, the adoption process of IDTV is situated within the "Diffusion of Innovations Theory" of Rogers (1995) that has been applied in various industries. For instance, Lawson-Borders (2003) used it to explain the adoption of new media by media companies. The model states that there are four phases preceding the actual adoption of an innovation, namely, (1) awareness, (2) interest, (3) evaluation, and (4) trial. Consequently, we try to formulate an answer to the following research questions:

- **Knowledge:** What is the knowledge of advertising professionals concerning IDTV and advertising on IDTV? Are they aware of the introduction and the possibilities of this new medium?

- **Attitudes:** How do advertising professionals perceive IDTV as a marketing communication tool? What are their attitudes toward IDTV?

- **Behavioral intentions:** Are advertisers willing to adopt IDTV in their marketing mix? Do they want to invest in IDTV?

- Is there a significant relationship between knowledge and attitude components on the one hand and the intention to integrate IDTV advertising in the marketing mix on the other?

Background

Definition of IDTV

IDTV can be defined in several ways, accentuating different attributes or technologies. Some examples are two-way TV (Jensen & Toscan, 1999), WebTV, one TV, enhanced TV, set-top-box TV (Van Stelten, 2004), digital TV, IPTV, and so on. Yet, no definition of this medium/new technology is widely accepted (Carey, 1997; Steuer, 1992; Stewart, 1999) and even at present, the medium has received little academic attention (Kang, 2002; Kim & Sawhney, 2002). Table 1 focuses on the main characteristics of *traditional TV* compared to IDTV. In this chapter IDTV will be defined from a consumer point of view: "IDTV is a group of technologies (1) that gives users (2) the possibility to take control (3) over their domestic TV experience (4), enabling interactivity with the content (5)." (1) IDTV is a group of technologies, among which the PVR, VoD, interactive program guide (IPG), and EPG, rather than a medium on its own. There are also different transmission channels. (2) The viewer is replaced by the user, who can use his/her TV for all kinds of activities, for example, banking, e-mailing, gaming, watching TV, or playing along in a game. All this by making use of his/her domestic TV (compared to the PC as end device). (3) The user is not obligated to take control over his/her TV; he/she can still use the TV in the traditional way. However, the user can choose what he/she wants to see, whenever he/she wants to see it by employing the technologies like the PVR and VoD. (4) Using TV will become an experience that extends traditional TV viewing. (5) Interactivity consists of three constructs (McMillan & Hwang, 2002): *User control*. The user search for additional information, e.g., for programs, VoD; *Two-way communication*. The user can use IDTV to send e-mails, play along in a game, chat on-screen, and so forth; *Synchronization:* The interaction and the feedback are not necessarily simultaneous in all cases. When playing along or gaming, the interaction has to be synchronic, while other activities such as e-mailing and VoD are less sensitive to time delays. Since in the empirical study, besides knowledge of IDTV advertising by advertising professionals, also attitudes, perceptions, and intentions will be explored, in the next subsections the potential advantages and disadvantages of IDTV advertising and the possible reactions of advertising professionals are highlighted.

IDTV Threats for Advertisers

First, the digitalization of TV entails the introduction of a great amount of new TV channels, which gives the TV viewer a much wider choice. For advertisers this implies a decrease in terms of reach per channel (O'Connor & Galvin, 2001). Van Den Broeck, Pierson, and Pauwels (2004) nuance this by stressing that the majority of the population only watches a limited number of channels.

A second threat comes along with the introduction of associated technologies of IDTV such as the PVR and VoD, which are that redefine the linearity of the program. Fortunato and Windels (2005) describe the functionality of the PVR as follows: *"The PVR allows users to record and store programming digitally, skip commercials with a touch on a button, pause live TV at any time, record programming through a digital menu, ..."* (p. 141). VoD, on the other hand, permits the user to choose his favorite program or movie from the offer of the broadcaster, enabling him/her to schedule his/her own TV evening (Wirtz & Schwartz, 2001). Both the PVR and VoD give more control to the viewer (Van Dijck, Heuvelman, & Peters, 2003). Time shifting, the possibility to pause, stop, and watch the program whenever it is convenient for the viewer, is an important feature of the PVR and VoD. Media planning based on television viewing figures and scheduling of programs are threatened by these new viewing patterns. Overall, the role of the broadcasters, and thereby advertisers, in the value chain is threatened to be replaced by the end user, choosing his/her own content at the time he/she wants to. Additionally, both technologies make it very

Table 1. Characteristics of traditional TV and IDTV

Traditional TV	Interactive, digital TV
Average picture quality	High picture quality
Average amount of channels	High amount of channels
One-way communication	Two-way communication
One-to-many	Many-to-many
Passive viewer	Active user
Ad- Interruption model	Ad-Permission model
Puch model	Pull model
Advertising driven revenu model	Commerce driven revenu model
Time restricted (Lineaire	Time-shifting (Participatory
Device centric	Ubiquitous
Entertainment	Entertainment, shopping
Static content	Dynamic content (updated constantly)
Average community building	High community building
Broadcasting	Narrowcasting
Public good - free	Pay TV
National	Global
TV rates through panels	Actual TV rates through STB
Ad zapping	Ad skipping
VCR	PVR
Domesticated	Penetration still low

easy for users to skip commercials (Boddy, 2004; Thawani, Gopalan, & Sridhar, 2004). Studies reveal that consumers who own a PVR skip up to 88% of all ads (Forrester, 2005). When comparing these figures with a recent study in The Netherlands showing that 20% of the respondents zap away commercials on analogue TV and 43% sit before the TV but do not watch, the PVR forms a real threat (Nederlandse Bond Van Adverteerders, 2005). Mediaedge (presentation by Fletcher, 2005) puts the skipping figures of the PVR in perspective. Their calculations show that in 2010 and under the assumption that 30% of the households own a PVR, the total available audience will only decrease by about 9%. A fourth threshold for advertisers can be the limited adoption of IDTV by the end user that is very important in terms of reach. To convince people to make the switch from TV as a public good to pay for IDTV, marketers are trying to find out what could be the relative advantage (Sarrina Li, 2004) or the killer application of IDTV. Napoli (2001) points out a fifth possible risk for advertisers not to adopt IDTV. He claims that advertisers are shifting communication budgets away from advertising-supported media to other marketing venues such as direct response mail and event sponsoring, as a result of the declining quality of the audience measurement induced by the enlarged inter- and intra-media choice and the declining willingness of respondents to participate in audience panels. As a result of these threats and risks, the question arises if advertisers will hesitate to invest in TV advertising in the future? Will the traditional TV business model based on advertising budgets fall apart?

IDTV Opportunities for Advertisers

IDTV has often been conceptualized as the convergence of traditional TV and the Internet. This implies that IDTV can build on the strengths of both media as an advertising channel as demonstrated in Table 2. When comparing the characteristics of advertising on IDTV with traditional commercials, the opportunities for advertisers boil down to two categories, that is, the development of personalized TV and the development of new formats.

Personalized TV

Through the increase of the number of channels via digitalization of TV, most channels will offer more specific content, which leads advertisers to believe that they can target their viewers more specifically. Contrary to what is often claimed, Deighton and Barwise (2000) assume that viewers of these new channels will not be as segmented as the readers of magazines. Although the content of the channels will be specific, this does not imply the public will be. Yet, IDTV offers the unique possibility to contact the target group very narrowly and the concept van addressability

Table 2. Characteristics of traditional TV, Internet, and IDTV as an advertising medium

Traditional TV	Internet	IDTV
Relatively passive	Relatively interactive	Continuum passive – interactive
High penetration (98% in EU)	Low penetration (37.4% in EU)	Potential high penetration (in the future)
Reach: national, broad	Reach: global, specific	Reach: global, specific
High emotional impact	Low emotional impact	High emotional impact
High involvement	High involvement	High involvement
Domesticated	Requiring a degree of computer literacy	High accessibility
Limited amount of information	High amount of information	High amount of information
Ad processing time	No time limit	No time limit
Suited for branding	Suited for direct marketing (branding)	Branding + direct marketing
No possibility to collect consumer data	Possibility to collect consumer data	Possibility to collect consumer data
Difficult to measure the	High accountability	High accountability
Push model P	ull model	Push & pull model
High ad irritation A	verage ad irritation	Average ad irritation

(Deighton & Blattberg, 1991) is precise to its place in this IDTV context. The PVR automatically saves all the programs the family watches, making it possible through data mining techniques to draft a viewer profile of each family unit, enabling narrowcasting of personal commercials to the individual set-top box (Chorianopoulos, Lekakos, & Pramataris, 2001; Gal-Or & Gal-Or, 2005; Lekakos & Giaglis, 2004). This IDTV application is called personalized TV or one TV (Wirtz & Schwartz, 2001); *"Personalization in interactive and future TV aims at targeting content to individual users by adapting the content based on user's likes and circumstances."* (Thawani, Golapan, & Sridhar, 2004, p. 1).

The commercial is transmitted and saved on the set-top box, decreasing the transmission cost. This data collecting set-top box is also very attractive from an overall customer relationship management perspective (Peltier, Schibrowsky, & Schultz, 2003). An important issue with personalized TV is the debate concerning consumer privacy, in which the regulatory authority will play a significant role, threatening the overall audience measurements, as discussed earlier.

New Formats

The development of digital television also brings along the possibility to develop new advertising formats. Looking at the uses and gratifications of traditional TV, Internet, and IDTV by consumers, (Livaditi, Vassilopoulou, Lougos, & Chorianopoulos, 2003; Morrison & Krugman, 2001), it appears that the convergence of TV and the Internet will not be complete (Coffey & Stipp, 1997). Concluding from earlier empirical consumer research, traditional TV will still be best suited to use for relaxing and passing time (Lee & Lee, 1995), where the Internet is still best suited for the acquisition of information, certainly in a business environment (Ferguson & Perse, 2000, 2004). Practical evidence is provided by The American Research Center Forrester (2005) who state that people who own a PVR still watch real life (traditional, linear) TV about 40% of their time. As a result of the fact that the traditional viewing will not disappear in the near future, the 30" commercial will also stay on screen, but maybe not in its current form. Interactive advertising will no longer be a concept only usable on the Internet, but will become applicable on television as well. Stewart (2004) defines interactive advertising as follows, "... *the presentation of information through mediated means, whether a computer or a mobile phone, and mutual, relative immediately interaction between consumers and marketers*" (p. 10). This implies a control shift from the advertiser to the consumer (Kitchen, 2003), who can decide to gather and process information of any kind. TV advertising shifts from a push to a pull philosophy. The consumer's motivation to interact with commercials becomes a key variable to comprehend the effects of this kind of marketing communication (Stewart, 2004). From this control shift arises the advertiser's opportunity to build a dialogue with his/her consumers (Godin, 1999). Furthermore, research has pointed out that interactivity in ads can enlarge the memory and attitude effect (Fortin & Dholoakia, 2005). Table 3 gives an overview of the new IDTV ad formats that are being developed.

In the Broadcast Stream

The Dedicated Advertising Location (DAL, mini-DAL) consists of a "*30-second TV ad embedded with clickable content 'micro sites' featuring individual still screen providing additional information*" (Bellman, 2004, p.2). The viewer can navigate through the extra information, thereby increasing the actively involved time, enlarging the cognitive processing of the advertising message, and as a result extending the advertising effectiveness. For instance, a study by Bellman (2004) demonstrated that the effects of three exposures to a traditional analogue commercial are equivalent to one exposure to a DAL, assuming the consumer interacts with this DAL. In the *impulse response format,* the extra information is placed on top of the screen like an Internet banner (Body, 2004) and is mostly used for direct marketing purposes (to order coupons, samples, or brochures) (interactive digital sales [IDS], 2005). It focuses on the impulse behavior that new media try to provoke. An important difference between the impulse response format and the DAL is that, in the former, the viewer stays in the linear *broadcast stream*. With the DAL, on the other hand, if the viewer presses the red button that appears in the commercial in the linear broadcast stream, he/she enters a second (broadband) stream, where the extra information is shown, similar to an Internet site. As a result, the viewer will miss a part of the program. With this problem in mind, Lekakos, Papakiriakopoulos, and Chorianopoulos (2001) introduced the *micro site*. A picture-in-picture or split screen divides the television screen in two parts and allows the user to navigate through the advertising content without missing part of the linear broadcasting stream. Yet, because the attention of the viewer is divided in two simultaneous streams, this format may lead to less effective advertising. Another solution for the user is to save the program or advertisement on the PVR through a quick push on the "bookmark" button, allowing the viewer to navigate through the information at his/her convenience (Lekakos et al., 2001). Another new format, not resulting in content overlap, is the *contact me* function that appears during the commercial. With a press on the red button, the user can choose to be contacted by the advertiser. Besides using the new formats, advertisers also have to cope with higher skipping rates of commercials through the introduction of the PVR and VoD. Advertising messages that are skip-proof, in other words *embedded in the television content*, for example, product placement, sponsorship, virtual advertising, on-screen banner advertisement during programs, and advertainment (a program totally made by an advertiser) (Body, 2004) will gain importance in the coming years (Saatchi & Saatchi Quattro, 2005). Digimedia uses the figures of MediaPost (The Media Center at the American Press, 2004) to demonstrate that the American investments in "branded entertainment" increased by 50% last year. New variants of product placement, like interactive product placement, where the user clicks on a button appearing in the program, bringing him/her to the DAL of the advertiser, are likely to emerge. Virtual advertising and sponsorship uses image manipulation, made possible by the digitalization of television (Body, 2004), for example, virtual billboards at live sports events or in news items. The use of logos and banners, adopted from the Internet, will also become visible on television.

Besides the Broadcast Stream

Alongside the broadcasting stream, IDTV allows advertisers to reach their target audiences through banners and logos on the EPG. Besides this, the broadcaster's portal can also be used to post advertising messages. In the "walled garden," advertisers can become visible through sites, games, gambling, chatting, and so forth. Consumers with IDTV technology can receive e-mails on their televisions. This opens up new opportunities for e-mail marketing. Finally, VoD allows the user to choose a movie, soap, or other content and is a possible way to let long commercials or advertainments be pulled by the user; a challenge for advertising professionals to make their commercials as attractive as possible.

Reactions of Advertisers

In terms of behavior, advertisers have two options to react to the development of IDTV. They can either reallocate their advertising budget to other media than television to avoid the threats posed by IDTV, or they can invest in the new advertising formats to take advantages of the opportunities IDTV has to offer (American Advertising Federation [AAF] Survey, 2004).

The trend to withdraw advertising budgets from TV is not new (Kim, Han, & Schultz, 2004). The customization of products has changed marketing from a mass to a one-to-one approach, where direct personalized marketing has become more and more important (Korgaonkar, Karson, & Akaah, 1997). Besides the growing attention for personalized communication at the expense of mass marketing, TV advertising specifically has an extra problem, namely, the increasing ad clutter that makes the effectiveness of TV commercials doubtful (Lowrey, Shrum, & McCarty, in press).

On the other hand, IDTV can make TV advertising more appealing for the consumer as well as for the advertiser. The consumer will only receive commercials and additional information that are of interest to him or her, decreasing the irritation and increasing the attention for the commercials, which results in an increased advertising effectiveness. From a historical perspective Galbi (2001) pointed out in his study that new media technology will probably not generate rapid growth in advertising revenues for media industries as a whole. A shift in advertising budgets are more likely to appear. In the U.S. the advent of radio and TV shifted about half of the print advertising share to these new media.

Table 3. IDTV advertising formats

In the broadcast stream	Besides the broadcast stream
In AD	**Walled garden**
* DAL (Mini-DAL)	* Logos/banners
* Impulse response	* Games
* Microsite	* Websites
* Contact me	**EPG/IPG**
In the content	* Logos/banners
* Product placement	**Direct Mailing (email)**
Sponsoring	**VOD**
Advertainment	* Long commercials
Interact.prod.placement	* Advertainment
Virtual advertising	
* On screen banners/logos	

Empirical Results

Research Method and Data Collection

The main objective of the present study is to explore the knowledge; the attitudes and perceptions; and the behavioral intentions towards IDTV as a marketing communication tool, of Belgian advertising professionals. A questionnaire was developed, based on previous studies (AAF, 2004; Bush, Bush, & Harris, 1998; Hsu, Murphy, & Purchase, 2001; Katz, 2004; Lace, 2004; Rodgers & Chen, 2002; Weapon7, 2002) and on depth-interviews with experts in the field of IDTV (Ogilvy; I-merge; Zappware; GfK; VMMa) (Ducoffe, Sandler, & Secunda, 1996). The questionnaire was pre-tested among advertisers and staff of advertising agencies in Belgium. The final questionnaire was organized around the most important phases of Rogers's diffusion theory (1995): (1) knowledge of IDTV and interest; (2) evaluation of IDTV as a marketing communication tool; and (3) intentions to invest in IDTV advertising (Trail & Adoption). With respect to the knowledge of IDTV and interest, general and specific questions concerning IDTV and advertising possibilities on IDTV were asked. The evaluation of IDTV was measured through questions concerning perceived advantages and disadvantages, objectives to use IDTV, and so forth. Questions regarding behavioral intentions focused on the intentions to try and adopt IDTV in the marketing communication mix and to invest in it. Because Dutch and French are Belgium's most common official languages, the questionnaire was offered in the two languages. The sampling frame was a database of 6,548 marketing professionals, all staff of advertising companies or communication consultancy agencies (Insites Consultancy and Pub Magazine). The URL of the Web survey was mailed to each

Table 4. Characteristics of the sample

Traditional TV	Interactive, Digital TV
Average picture quality	High picture quality
Average amount of channels	High amount of channels
One way communication	Two way communication
One-to-many	Many-to-many
Passive viewer	Active user
Ad-interruption model	Ad-permission model
Push model	Pull model
Advertising driven revenue model	Commerce driven revenue model
Time restricted	Time shifting
Device centric	Ubiquitous
Entertainment	Entertainment, shopping, email, gaming, ...
Static content	Dynamic content (updated constantly)
Average community building	High community building
Broadcasting	Narrowcasting
Public good - free	Pay TV
National	Global
TV rates through panels	Actual TV rates through STB
Ad zapping	Ad skipping
VCR	PVR
Domesticated	Penetration still low

member of the sampling frame. The response rate was 4.9%, in other words 320 respondents participated. This low response rate is normal for online questionnaires (Sheehan, 2002). Table 4 gives an overview of the most important characteristics of the sample. The questionnaire was sent out in the period just before the commercial introduction of IDTV in Belgium in May 2005.

Results

Knowledge of IDTV Advertising

The general awareness of advertising professionals about the possibilities of IDTV is very weak (Figure 1). Only 17.8% of the respondents declared to have a good to very good knowledge of the possibilities of IDTV, compared to 60.9% stating to have a weak to very weak knowledge of IDTV. When comparing the knowledge of advertisers with people working in agencies, it appeared that professionals in advertising agencies were better informed (18.6% assumed to have a good to very

Figure 1. General knowledge of IDTV

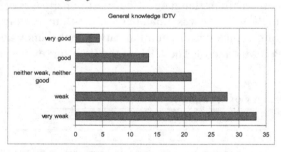

Figure 2. Knowledge of IDTV advertising possibilities

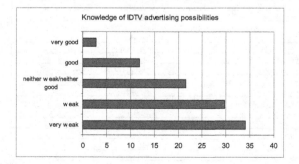

good knowledge and 55.6% said to have a weak to very weak knowledge), but the difference between the two groups was not significant (Chi-squared, p=0.204).

To the specific question in which country the adoption of IDTV is highest, 39.1% answered Japan, followed by 33.1% in the U.S. Only 14.3% of the respondents gave the right country, the UK. This indicates the low general knowledge of IDTV by the Belgian advertising world. In Europe, UK, Sweden, Ireland, France, and Denmark have the highest level IDTV household penetration; the U.S. and Japan are lagging behind (Ipsos, 2004). The acquaintance of the advertising possibilities on IDTV is even lower than the general knowledge as pointed out in Figure 2.

Merely 14.7% of the professionals maintained to have a good to very good knowledge about the advertising possibilities on IDTV, while about 64% of the respondents assumed to have a weak to very weak knowledge. There was no significant difference between the knowledge of advertisers and agencies for the advertising possibilities on IDTV.

Looking at the specific awareness of the advertising formats in Figure 3, the DAL (known by 19.3% of the respondents), mini-DAL (14.5%), and walled garden applications (22%) were known by only a small percentage of the professionals.

Figure 3. Knowledge of IDTV ad formats

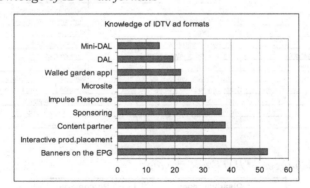

Table 5. Information needs about IDTV advertising

Information needs	Percentage
Cost	47,8%
Case studies	44,4%
ROI	44,1%
Available data	43,4%
Business implications	37,8%
Technology	30,6%
Ad formats	27,8%
Production proces	21,9%

Sponsoring (known by 55.4% of the respondents), banners on EPG (52.7%), and advertainment (37.9%) were known by more advertising professionals. This general lack of knowledge could be one of the major thresholds for advertiser professionals to invest in IDTV, as will be demonstrated later in this chapter.

With respect to one topic, namely advertising formats, advertising agencies showed a significantly (Chi-squared, p<0.001) higher need for information than advertisers. Of the professionals, 43.3% from agencies versus 14.7% of the advertisers would like to receive more information about these formats.

Attitudes Towards IDTV Advertising

With respect to the attitudes and perceptions of IDTV, 61.7% of the advertising professionals perceived IDTV more as a tool for action communication than for developing positive brand values with consumers; 19.7% stated that IDTV was

Table 6. Advantages of IDTV advertising

Traditional TV I	nternet	IDTV
Relatively passive R	elatively interactive	Continuum passive - interactive
High penetration (98% in EU)	Low penetration (37,4% in EU)	Potential high penetration (in the future)
Reach: national, broad	Reach: global, specific	Reach: global, specific
High emotional impact	Low emotional impact	High emotional impact
High involvement	High involvement	High involvement
Domesticated	Requires a degree of computer literacy	High accessibility
Limited amount of information	High amount of information	High amount of information
Limited ad processing time	No time limit	No time limit
Suited for branding	Suited for direct marketing & branding	Suited for direct marketing & branding
No possibility to collect consumer data	Possibilty to collect consumer data	Possibilty to collect consumer data
Difficult to measure the ROI	High accountability	High accountability
Push model	Pull model	Push & pull model
High ad irritation	Average ad irritation	Average ad irritation

Table 7. Disadvantages of IDTV advertising

Disadvantages	Percentage
Too little experience	55,3%
Adoption by end user	45,9%
Costs	38,4%
Too new	28,1%
Complex production proces	18,1%
No accurate measurements	17,5%
Justification of investement	15,0%
Differences in platform	11,9%
Technical problems	11,6%
Bad results	9,4%
IDTV-logistics	8,8%
Flexibility of the broadcaster	5,9%
Not effective	5,9%

suited for both goals and 18.8% perceived IDTV as more appropriate for branding purposes. This is similar to the evolution in the UK, where the first advertisers using IDTV were direct marketers. The respondents were asked to indicate their top three of the perceived advantages and disadvantages of IDTV advertising. The possibility to provide more product information (mentioned by 38.1% of the respondents), more specific targeting (35.6%), and two-way communication (35.0%) were the most frequently mentioned advantages, as pointed out in Table 6. Other aspects like the possibility to position the brand as innovative, increasing brand awareness, and generating sales were also perceived as strengths of advertising on IDTV.

Table 7 shows the most important perceived disadvantage was that IDTV is not yet a mainstream medium. This lack of general advertising experience was perceived as a problem by 55.3% of the respondents. The low adoption of IDTV by the end user was another perceived weakness of IDTV as a marketing communication tool (45.9%). The cost of advertising on IDTV was the third most frequently mentioned disadvantage (38.4%).

There were no significant differences between advertisers and agencies concerning the perceived advantages and disadvantages of IDTV advertising. Marketing professionals were convinced that IDTV would become a mainstream medium in the near future (51.1% agree) and believe that the convergence of TV and PC will take place (55.2% agree), but there was some doubt that interactivity would work for advertising (28.5% doubt). About half of the respondents (47.7%) believed that the 30 sec commercial would still exist in the near future, while 50.4% were afraid that PVR users would skip all commercials.

Intentions to Use IDTV Advertising

The respondents were asked if they were planning to invest in IDTV and adopt IDTV in their marketing communication mix. Figure 4 points out that 33.2% of

Figure 4. Intentions to adopt IDTV

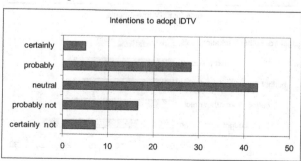

the marketing professionals, who did not have any experience with IDTV (only six respondents had some experience with IDTV), intended to adopt IDTV probably or certainly. The largest part, 42.9% took a neutral stand, and 23.9% of the professionals claimed not to be willing to invest in IDTV at all.

With respect to their specific interest in the IDTV ad formats, Figure 5 shows that 45.5% of the respondents were interested in the DAL, followed by the impulse response format (33.8%), and the interactive product placement possibilities (30.3%); formats that were not well known among the professionals, as mentioned earlier. Looking at the top three thresholds for adopting IDTV as a marketing communication tool, the most mentioned variable was the lack of general knowledge about IDTV (answered by 44%).

The low adoption of IDTV by the end user was mentioned by 21% of the respondents. Some respondents (16.7%) stated that IDTV was just not suited for their

Figure 5. Interest in IDTV ad formats

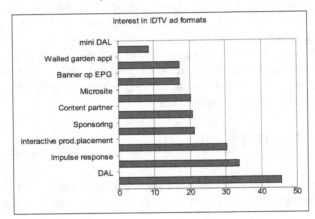

Figure 6. Source of IDTV investments

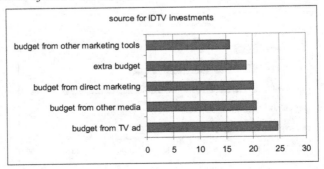

product. Key motivators for the respondents that would incite them to adopt IDTV were reducing the costs (56.3%), enlarging the adoption of IDTV by the end user (54.1%), and improving the technology (38.8%). An important question concerning the investment in IDTV as a marketing communication tool is the marketing budget that would be allocated to IDTV. Figure 6 shows that one quarter of the respondents would reallocate the TV advertising budget to use for IDTV advertising. Twenty-one percent of the respondents would reallocate budgets from other advertising media, and 20% stated to reallocate the direct marketing budget to use for IDTV advertising.

Relationship Between Knowledge and Attitudes and Intentions

It is interesting to look at those variables that are assumed to have an influence on behavioral intentions. Rogers' (1995) model states that awareness (of the product/ innovation) and positive attitudes towards it are prerequisites to behavioral intentions. Table 8 gives an overview of a number of significant correlations found. The knowledge of advertising professionals about the opportunities of IDTV was positively, significantly correlated with the intention to adopt IDTV in the marketing mix. Advertising professionals with more knowledge of IDTV had a stronger intention to adopt IDTV in the near future.

Looking at attitudes towards IDTV, we found a positive, significant relation between the perception that IDTV is the best medium to advertise and intention to adopt IDTV. Consistently, we also found a negative, significant relation between the idea that IDTV will only work for programs and not for advertising and intentions to adopt the medium for advertising goals. As mentioned earlier, most respondents (61.7%) perceived IDTV as a tool well suited for direct communications as opposed to branding goals. This implies that advertising professionals who emphasize the overall importance of direct marketing should have a higher intention to adopt IDTV

Table 8. Correlation values

Correlations (n= 301)	Spearman's Rho	p-value
Knowledge of IDTV Intention to adopt IDTV	0,259	<0,01
IDTV is the best ad medium Intention to adopt IDTV	0,232	<0,01
IDTV works for progr. not for ads Intention to adopt IDTV	-0,203	<0,01
Importance of direct marketing Intention to adopt IDTV	0,171	0,003
Invest more in IDTV then in TV Intention to adopt IDTV	0,232	<0,01
Approach by others Intention to adopt IDTV	0,226	<0,01

than advertisers who stress branding. The correlation between the intention to adopt IDTV and the importance attached to direct marketing was weak but significantly positive. These results fit the model of Rogers (1995) where a positive evaluation of the innovation influences the trial and adoption rate. Another important element of Rogers' theory is that interpersonal communication has an influence on the intention to adopt an innovation. Our data showed that only 10.3% of the respondents have been approached by a third party (e.g., broadcasters, advertising agencies, advertisers, telecommunications operators, consultants, and interactive agencies) concerning IDTV advertising possibilities. There was a weak but significant positive correlation between the variables "intention to adopt" and "*being approached or not*"). This implies that respondents being approached by intermediates had a stronger intention to adopt IDTV, compared with those not being approached.

Conclusion

The main objective of this chapter was to focus on the development of IDTV from the perspective of advertising professionals. As pointed out by Leckenby (2003), the success of a new medium depends on the adoption by advertisers. From a network perspective, the adoption by advertisers can be viewed as an indirect externality (Markus, 1990) to the total system, meaning that the investments of the advertising world will decrease the production costs, making the content more acceptable for end users adopting the medium. The adoption process of IDTV by advertisers can be placed within Rogers' (1995) theory of the diffusion of innovations. In the first phase, the awareness of and interest in the innovation has to be developed. The results of the present study reveal that at the start of the commercial introduction of the technology in Belgium, the advertisers' general knowledge of IDTV is still low. The specific knowledge concerning the new advertising formats is even lower. Most respondents mark this lack of knowledge as a threshold to invest in IDTV. The perceptions and attitudes towards IDTV can be placed in Rogers' (1995) "*phase of evaluation*" of the innovation. IDTV is a marketing tool usable for branding as well as for direct marketing goals. However, the majority of the Belgian advertising professionals perceive IDTV more as a direct marketing tool. Among the strengths of the medium, the respondents perceive the possibility to provide more product information, two-way communication with the consumers, and specific targeting. Perceived disadvantages are the low overall experience with IDTV, the low adoption of IDTV by the end user, and the high costs of advertising on IDTV. Another perceived disadvantage is the low penetration of IDTV by the end user in Belgium. This is evident, because the measurement took place before the commercial introduction. However, the penetration of IDTV in Belgium is predicted to be very slow because of the high amount of television channels (33 on average) already currently available to the Belgian viewer through cable (penetration 99.8% in Belgium). Also, there is no tradition of pay TV in Belgium. The low experience with IDTV in general is also a disadvantage for the medium, probably increasing the perceived risk of adopting IDTV. The results of the intentions to invest in IDTV, taken into account the lack of knowledge and the fact that some of the respondents use no above-the-line media, are promising. Only one quarter of the respondents stated not to intend to invest in IDTV in the future, while the majority of advertising professionals had not yet developed an opinion concerning their IDTV investments.

Implications and Future Research

This study has a number of implications. Overall, the respondents believe that IDTV will become a mainstream medium in the future. A substantial proportion of the Belgian marketing professionals appear to be interested in integrating IDTV advertising in their marketing communication mix. Interpersonal communication between advertising professionals and people from the broadcasting companies or telecommunications operator may play an important role in this phase. Because the perception of the usability of IDTV has an influence on the profile of the professional's intention to adopt, it is important to stress the possibilities of IDTV for branding as well. This will ensure that advertising professionals valuing branding will become potential users of IDTV.

Future research could focus on the media investment behavior of advertisers. Will they actually invest extra budgets in IDTV or will they reallocate their TV budgets to other media? For media, shifts in advertising budgets are important to foresee to be able to search for other, additional revenue streams. How does this trend evolve in other countries? This research question could be tackled by using the framework of the *niche theory* (Dimmick, 1997), assuming limited media resources and focusing on uses and gratifications research of consumers. Another relevant track for further research is the investigation of variables influencing the adoption process of IDTV by an advertiser, from a diffusion of innovation perspective. These research questions are important for broadcasters and other parties in the IDTV value chain to lower the thresholds for advertisers to adopt IDTV.

References

American Advertising Federation (AAF). (2004). *Survey of industry leaders on advertising trends*. Washington, DC: Atlantic Media Company.

Bellman, S. (2004). *The impact of adding additional information to television advertising on elaboration, recall, persuasion.* Paper presented at the ANZMAC Conference, Wellingthon.

Boddy, W. (2004). *New media and popular imagination.* UK: Oxford University Press.

Bush, A. J., Bush, V., & Harris, S. (1998). Advertiser perceptions of the Internet as a marketing communications tool. *Journal of Advertising Research, 38*(2), 17-28.

Carey, J. (1997). Interactive television trials and marketplace experiences. *Multimedia Tools and Applications, 5*(2), 207-216.

Chorianopoulos, K., Lekakos, G., & Pramataris, K. (2001, June). *Metrics for advertisement effectiveness measurement in the interactive TV environment: The iMedia case.* Paper presented at the 14th Bled electronic commerce conference, Slovenia.

Coffey, S., & Stipp, H. (1997). The interactions between computer and television usage. *Journal of Advertising Research, 37*(2), 61-67.

Deighton, J., & Barwise, J. (2000). *Digital marketing comunication.* Working Paper. London Business School, Future Media Center.

Deighton, J. A., & Blattberg, R. C. (1991). Interactive marketing: Exploiting the age of addressability. *Sloan Management Review, 33*, 5-14.

Dennis, E. E. (2002). Prospects of big idea—Is there a future for convergence? *International Journal of Media Management, 5*(1), 7-11.

Dimmick, J. (1997). The theory of the niche and spending on mass media: The case of the "video revolution." *Journal of Media Economics, 10*(3), 33-43.

Ducoffe, R. H., Sandler, D., & Secunda, E. (1996). A survey of senior agency, advertiser, and media executives on the future of advertising. *Journal of Current Issues and Research in Advertising, 18*(1), 1-19.

Ferguson, D. A., & Perse, E. M. (2000). The world wide Web as a functional alternative to television. *Journal of Broadcasting and Electronic Media, 44*(2), 155-174.

Ferguson, D. A., & Perse, E. M. (2004). Audience satisfaction among TiVo and replay users. *Journal of Interactive Advertising, 4*(2). Retrieved August 2005, from www.jiad.org/vol4/no2/ferguson

Fletcher, D. (2005, May). *ITV and the future.* Paper presented at the ITV congress, London.

Forrester Research. (2005). *Consumer technografics.* North American Study, US.

Fortin, D. R., & Dholoakia, R. R. (2005). Interactivity and vividness effects on social presence and involvement with Website-based advertisement. *Journal of Business Research, 58*(3), 387-396.

Fortunato, J. A., & Windels, D. M. (2005). Adoption of digital video recorders and advertising: Threats and opportunities. *Journal of Interactive Advertising, 6*(2). Retrieved November 2005, from www.jiad.org/vol6/no1/fortunato

Galbi, D. (2001). The new business significance of branding. *International Journal on Media Management, 3(*4), 192-198.

Gal-Or, E., & Gal-Or, M. (2005). Customized advertising via a common media distributor. *Marketing Science, 24*(2), 241-253.

Godin, S. (1999). *Permission marketing: Turning strangers into friends, and friends into customers.* London: Simon & Schuster.

Hsu, T. W., Murphy, J., & Purchase, S. (2001). Australian and Taiwanese advertiser's perceptions of Internet marketing. *Australian Marketing Journal, 9*(1), 33-45.

Interactive Digital Sales (IDS). (2005). *Interactive advertising.* Retrieved August 2005, from http://www.idigitalsales.co.uk/interactiveadvertising/

Ipsos. (2004). *News.* Retrieved February 2005, from http://www.ipsos.com/news/CorporateNews.aspx

Jensen, F. J., & Toscan, C. (1999). *Interactive TV. TV of the future or future of TV? Media and cultural studies 1.* Denmark: Aalborg University Press.

Kang, M. H. (2002). Interactivity in TV: Use and impact of an interactive program guide. *Journal of Broadcasting & Electronic Media, 46*(3), 330-345.

Katz, B. (2004). *Interactive TV survey 2004.* London: Interactive Digital Sales.

Kim, I., Han, D., & Schultz, D. E. (2004). Understanding the diffusion of integrated marketing communications. *Journal of Advertising Research, 44*(1), 31-45.

Kim, P., & Shawhney, H. (2002). Machine-like new medium—Theoretical examination of ITV. *New Media & Society, 47*(2), 217-233.

Kitchen, P. (2003). *Future of marketing.* New York: Pallgrave McMillan.

Korgaonkar, P. K., Karson, E. J., & Akaah, I. (1997). Direct marketing advertising: The assents, the dissents, and the ambivalents. *Journal of Advertising Research, 37*(5), 41-55.

Lace, J. M. (2004). At the cross road of marketing communications and the Internet: Experiences of UK advertisers. *Internet Research, 14*(3), 236-244.

Lawson-Borders, G. (2003). Integrating new media and old media. *International Journal On Media Management, 5*(2), 91-99.

Leckenby, J. D. (2003). *The interaction of traditional and new media.* Working Paper. University of Texas at Austin, Communication College.

Lee, B., & Lee, R. S. (1995). How and why people watch TV: Implications for the future of interactive television. *Journal of Advertising Research, 35*(6), 9-18.

Lekakos, G., & Giaglis, G. M. (2004). Lifestyle-based approach for delivering personalized advertisements in digital interactive TV. *Journal of Computer Mediated Communication, 9*(2).

Lekakos, G., Papakiriakopoulos, D., & Chorianopoulos, K. (2001). An integrated approach to interactive and personalized TV advertising, In M. Bauer, P. J. Gmytrasiewicz, & J. Vassileva (Eds), *User modeling 2001: 8th International Conference.* Berlin Heidelberg: Springer-Verlag.

Livaditi, J., Vassilopoulou, K., Lougos, C., & Chorianopoulos, K. (2003). Needs and gratifications for interactive TV applications: Implications for designers. *Paper presented at the 36th Hawaii International Conference on System Sciences, Hawaii.*

Lowrey, T. M., Shrum, L. J., & McCarty, J. A. (in press). The future of television advertising. In A. Kimmel (Ed.), *Marketing communication: Emerging trends and developments.* New York: Oxford University Press.

Maddox, L. W., & Mehta, D. (1997). The role and effect of Web addresses in advertising. *Journal of Advertising Research, 37*(2), 47-60.

Markus, L. (1990). Toward a critical mass theory of interactive media: Universal access, interdependence and diffusion, In J. Fulk & C. Steinfeld (Eds.). *Organizations and communication technology.* Newbury Park, CA: Sage.

McMillan, S. J., & Hwang, J. S. (2002). Measures of perceived interactivity: An exploration of the role of direction of communication, user control, and time in shaping perceptions of interactivity. *Journal of Advertising, 31*(3), 29-42.

Morrison, M., & Krugman, D. M. (2001). A look at mass and computer mediated technologies: Understanding the roles of television and computers in the home. *Journal of Broadcasting and Electronic media, 45*(2), 135-161.

Napoli, P. M. (2001). The audience product and the new media environment: Implications for the economics of media industries. *International Journal on Media Management, 3*(1).

Nederlandse Bond Van Adverteerders. (2005) *Schokkend nieuws voor de buis, registratie van het feitelijk kijkgedrag van consumenten naar reclame.* Research Study.

Negroponte, N. (1995). *Being digital.* London: Coronet Books.

O'Connor, J., & Galvin, E. (2001). *Marketing in the digital age.* London, New York: Prentice Hall.

Ofcom. (2005). *Digital TV UK update.* Retrieved September 2005, from http://www.ofcom.org.uk/media/news/2005/09/nr_20050915

Pagani, M. (2000). Interactive television: A model of analysis of business economic dynamics. *International Journal on Media Management, 2*(1), 25-36.

Peltier, W. J., Schibrowsky, J. A., & Schultz, D. E. (2003). Interactive integrated marketing communication: Combining the power of IMC, the new media and database marketing. *International Journal of Advertising, 22*(1), 93-115.

Rodgers, S., & Chen, Q. (2002). Post-adoption attitudes to advertising on the Web. *Journal of Advertising Research, 45*(3), 95-104.

Rogers, E. M. (1995). *Diffusion of innovations.* New York: Free Press.

Saatchi & Saatchi Quatro. (2005). *What's new? IDTV for advertiser.* Retrieved October 2005, from http://whatsnewsaatchi.blogspot.com/

Sarrina Li, S. C. (2004). Exploring the factors influencing the adoption of interactive cable television in Taiwan. *Journal of Broadcasting & Electronic Media, 48*(3), 466-483.

Sheehan, K. B. (2002). Online research methodology: Reflections and speculations. *Journal of Interactive Advertising, 3*(*1*). Retrieved October 31, 2005, from http://jiad.org/vol3/no1/sheehan

Steuer, J. (1992). Defining virtual reality: Dimensions determining telepresence. *Journal of communication, 42*(4), 23-72.

Stewart, D. W. (2004). The new face of interactive advertising: It's time to rethink traditional ad research strategy. *Marketing research, 16*(1), 10-15.

Stewart, J. (1999). Interactive TV at home: TV meets the Internet. In J. Jensen & C. Toscan (Eds.), *Interactive television, TV of the future or the future of TV?* Denmark: Aalborg University Press.

Thawani, A., Gopalan, S., & Sridhar, V. (2004). *Context aware personalized Ad insertion in an interactive TV environment.* Paper presented at the workshop of Personalization in Future TV, Eindhoven, The Netherlands.

The Media Center at the American Press. (2004). *Branded entertainment.* Retrieved October 2005, from http://www.mediacenter.org/content/3673.cfm

Van Den Broeck, W., Pierson, J., & Pauwels, C. (2004). *Does interactive TV imply new uses? Flemisch Case study.* Paper presented at the IDTV Conference, Brighton, UK.

Van Dijk, J., Heuvelman, A., & Peters, O. (2003). *Interactive television or enhanced television? The Dutch users interest in applications of ITV via set-top boxes.* Paper presented at the 2003 Annual Conference of the International Communication Association, Amsterdam.

Van Stelten, N. (2004). *In het Land der Blinden is Televisie Koning, E-view, 2.* Retrieved August 2005, from http://comcom.uvt.nl/e-view/04-2/inhoud.htm

Weapon 7. (2002). *ITV—A view from the trenches.* Research Study. UK.

Wirtz, W., & Schwartz, J. (2001). Strategic implications of the segment of one TV. *International Journal on Media Management, 3*(1), 15-25.

Chapter XVIII

Business Models with the Development of Digital iTV Services:
Exploring the Potential of the Next Transaction Market

Margherita Pagani, Bocconi University, Italy

Abstract

The advent of digitalization is providing big opportunities, which are changing the shape of the broadcasting industry. New business models and revenue opportunities based on digital capabilities are emerging. The purpose of this chapter is to outline the different business models adopted in Europe in terms of contents offered and related revenue opportunities. After reviewing the business model literature and analyzing the value curve of interactive television (iTV) services the chapter addresses the following research questions: (1) How to cross the chasm of knowledge? (2) How to explore the opportunities opened by new technologies? (3) Which trends will influence the launch of new ITV services? The chapter describes revenue flows among the value chain elements, critical success factors for achieving competitive position, the role of content, and customer gate keeping in the new competitive environment.

Introduction

The advent of digitalization is providing big opportunities, which are changing the shape of the broadcasting industry. New business models and revenue opportunities based on digital capabilities are emerging.

Digital technology provides digital media companies with the opportunity to:

- Provide new products and cross-media experiences aimed at individual consumers and like-minded groups;
- Provide premium services that give viewers access to greater and more personalized content;
- Change the television advertising model from mass media to targeted advertising (market and advertise on one to one basis direct to consumer); and
- Expand from a traditional push model to a pull model for product distribution.

Different business models have been proposed for monetizing interactive digital content. Revenue models are the key to effectiveness and sustainability.

The success of the interactive digital television (IDTV) industry depends on the development of sustainable revenue models that support value-adding companies at all stages of the value chain.

The value chain structure is now well established in the industry and will only get more solid as the industry matures.

The advantages of carrier portals as a mechanism for the provision and distribution of information and entertainment and the resulting benefits for the different players along the value chain is still being debated across the industry.

These deployments will bring existing TV programs to mobile handsets and will encourage the development of new iTV services. All the players in the market are facing many commercial challenges, particularly related to business models, intellectual property rights, spectrum allocations, and partnership arrangements.

The purpose of this chapter is to outline the different business models adopted in Europe in terms of contents offered and related revenue opportunities.

After reviewing the business model literature and analyzing the value curve of iTV services, the chapter addresses the following research questions: (1) How to cross the chasm of knowledge? (2) How to explore the opportunities opened by new technologies? (3) Which trends will influence the launch of new iTV services?

The chapter describes revenue flows among the value chain elements, critical success factors for achieving competitive position, the role of content, and customer gate keeping in the new competitive environment.

Business Model: The Theoretical Background

In order to provide a theoretical background, in this section we investigate the business model literature to understand what a business model is, along with various components that make up business models, and to understand the various uses of business models.

There are various definitions in the business literature of what constitutes a generic business model but some fail to pay explicit attention to technology (Weill & Vitale, 2001), while others fall short in the area of defining the multiplicity of actors. An appropriate definition proposed by Timmers (1998) lays emphasis on the architectural and technology elements.

Timmers defines business models as follows:

- An architecture for the product, service, and information flows
- Description of the various business actors and their roles
- A description of the potential benefits for the various business actors
- A description of the sources of revenues

Another definition of a business model suitable for the purpose of this research is proposed by Pigneur (2000) who defines a business model as follows:

- The architecture of an organization and its network of partners for creating, marketing, and delivering value
- Relationship capital to one or several segments of customers in order to generate profitable and sustainable revenue streams

Some researchers from the e-business and Internet arena focus on the revenue aspects of a business model (Rappa, 2001), other approaches look at the business model from the business actor perspective (Timmers, 1998). Afuah and Tucci (2001) introduce an approach to business models that emphasizes the value perspective and considers the creation of value through several actors. Business models can also be defined from network components perspective. Amit and Zott (2001) describe an e-business model as the architectural configuration of the components of transactions designed to exploit business opportunities. The approach by Petrovic, Kittl, and Teksten (2001) proposes breaking down a business model into seven submodels for describing the business processes that generate value. These submodels are the value model, the resource model, the production model, the customer relations model, the revenue model, the capital model, and the market model.

Schmid (2000a, 2000b) argues that we are facing a new industrialization and that in the digital economy the scarce resource shifts from production to communication in a novel way and, therefore, the entire design of value creation systems is challenged.

Westland and Clark (1999) elaborate the shift from a traditional business model for marketing to a new interactive electronic commerce business model.

Moore (1998) believes that the transition from the multidimensional form (M-form) to the ecosystem form (E-form) will be at the heart of future success and growth. A similar, more detailed model is presented by Österle (2000) who defines an intermediary that supports the entire customer process (process portal provider) using a variety of standardized electronic services.

In this study we adopt the approach suggested by Afuah and Tucci (2001) which emphasizes the value perspective and describes the business models according to the definitions provided by Timmers (1998) and Pigneur (2000). In the following paragraphs, we first describe the value of iTV services for users then we focus on the specific services and the network of partners for creating, marketing, and delivering value.

The Value Curve of Digital iTV Services

The term *interactivity* is usually taken to mean the chance for interactive communication among subjects. Technically, interactivity implies the presence of a return channel in the communication system, going from the user to the source of information (Pagani, 2000, 2003). The channel is a vehicle for the data bytes that represent the choices or reactions of the user (input).

This definition classifies systems according to whether they are diffusive or interactive.

Diffusive systems are those that only have one channel that runs from the information source to the user (this is known as *downstream*).

Interactive systems have a return channel from the user to the information source (this is known as *upstream*).

The first stage of our analysis focuses on the consumer and aims to identify a hierarchy of importance of factors influencing adoption of IDTV services and to describe the key descriptive elements of homogeneous segments of the population.

This stage of the analysis concentrates specifically on the main results emerging from a quantitative marketing research conducted in Italy in 2005 through phone questionnaires on a sample of 600 Italian TV viewers.[1] We consider Italy as it is one of the countries in Europe facing the migration process towards digital television (DTV).

The main goal of this stage of analysis is to understand the value curve in order to define the correct strategies for offering the new iTV services in a proper and differentiated way. To realize this research objective, conjoint analysis was seen as the appropriate statistical tool.

Sample and Methodology

A random sample of 600 Italian TV viewers was interviewed by phone in Italy in 2005. The demographics of the sample can be summarized as follows: gender (female: 51%; male: 49%); age (18-24 years: 13%, 25-34 years: 21%, 35-44 years: 19%, 45-54 years: 17%, 55-64 years: 16%, beyond 65 years: 14%), and geographical area (northwest 27%, northeast 19%, center 20%, south and islands 34%).

The methodology adopted is the conjoint analysis technique that allows a set of overall responses of factorially designed stimuli to be decomposed so that the utility of each stimulus attribute can be inferred from the respondents' overall evaluations of the stimuli (Green, Helsen, & Shandler, 1988). A number of (hypothetical) combinations of service elements can be formulated that can be presented to a sample of customers.

We used Adaptive Conjoint Analysis (ACA) to conduct our conjoint study. ACA is a PC-based system for conjoint analysis. The term *adaptive* refers to the fact that the computer-administered interview is customized for each respondent. At each step, previous answers are used to decide which question should be asked next to obtain most of the information about the respondent's preferences. The program allows the researcher to design a computer-interactive interview and administer the interview to respondents. The interview can consider many factors and levels, paying special attention to those the respondent considers most important.

Questioning is done in an "intelligent" way, the respondent's utilities are continually re-estimated as the interview progresses, and each question is chosen to provide the greatest amount of additional information, given what is already known about the respondent's preferences.

The key items considered in the questionnaire are as follows:

1. **Interest** for the service categories under scrutiny
2. **Preference** for means/platforms through which selected services can be accessed (portables, phone, and/or TV)
3. Analysis of **critical factors** influencing adoption of DTV
5. Ranking of **service features**

The Value Curve

We focus now only on results emerging from the survey, which are related to the value curve from the user perspective.

Each respondent was asked to indicate the most important features influencing his/her adoption of DTV.

As mentioned earlier, conjoint analysis allows us to define a hierarchy of importance concerning the critical factors influencing adoption of DTV.

It emerges that each group segmented by age perceives a different relative value for each component of the value proposition. We can distinguish specific value curves (Figure 1).

If we compare preferences expressed by people, we can see that usefulness is considered more important by people aged 35-44, price is more appreciated by people aged 25-34, and reliability of contents by young people (18-24). These differences are statistically significant.

Usefulness, ease of use, and price are considered the most important variables to access segments of the population, and this result confirms findings emerging from

Figure 1. The value curve of DTV (max 2 answers) – base 600 respondents. Source: New Media&Tv-lab, Bocconi University, 2005

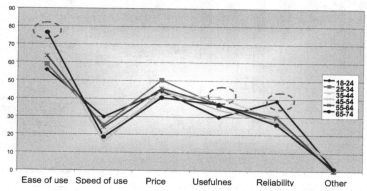

Table 1. Level of interest towards a sample of iTV services: Segmented by age

	18-24	25-34	35-44	45-54	55-64	65-74
Electronic Program Guide	5,0	4,0	3,9	3,6	3,3	3,4
Games	4,1	3,4	2,8	2,3	2,3	2,3
T-shopping	4,6	3,8	3,3	2,7	2,3	2,5
On-line banking	4,5	4,2	4,2	3,0	2,8	2,3
News	6,1	5,2	4,9	4,5	4,1	3,7
Financial news	4,1	3,8	3,6	3,2	2,8	2,4
Comunication	5,9	5,4	5,0	4,0	3,9	3,2
Music on demand	5,7	5,1	4,1	3,3	2,9	2,9
Maps	5,8	5,3	4,7	3,9	3,3	3,0
Educational	5,3	5,3	5,1	4,3	4,0	3,7
Voting	3,6	3,5	3,2	2,5	2,4	2,3
Betting	3,1	3,1	2,4	2,2	1,9	2,2
Local news	5,2	5,0	4,8	3,8	3,3	3,0

the literature of technology adoption acceptance (Davis, 1989; Davis, Bagozzi, & Warshaw, 1989).

Previous studies (Pagani, 2004) demonstrated that the variable with the greatest influence on people's behavior is the degree of interest toward the innovative services. We measure the level of interest for different categories of iTV services as shown in Table 1.

Many respondents state not to know the opportunities offered by the development of DTV and the knowledge is correlated with intention to adopt.

Lack of knowledge emerges as one of the most important barriers influencing adoption. Only 79.5% of respondents declare to know digital terrestrial television (DTT) and knowledge decreases with the age of respondents.

Figure 2. Knowledge of IDTV services in Italy—Base 600 respondents

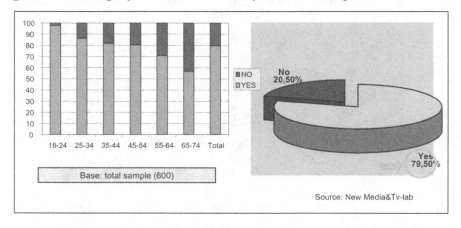

Next, it is apparent, as we have already found on the Internet, that understanding consumer behavior in interactive "channels" is crucial for fashioning interactive services. Knowing not only who is interacting but also when, where, and in what way they engaged, is invaluable. All of this, of course, is directly linked to determining what is the best way to get consumers to respond to the specific interactive service and interact with the brand.

Crossing the Chasm in the iTV Adoption

Future demand and penetration of iTV services is expected to grow very fast but knowledge emerges as one of the most important drivers influencing the interest and related adoption. The first critical issue that players involved need to address is the "crossing the chasm" paradigm (Moore, 1990) in the technology adoption life cycle model. Crossing the chasm is jumping into that empty area between the innovators segment and the early majority. The early adopters are iTV enthusiasts and are always looking forward to experience technology innovations. Being the first, they are also prepared to bear with the inevitable bugs that accompany any innovation just coming to market. By contrast, the early majority is looking to minimize the discontinuity with the old ways, and they do not want to debug somebody else's product. By the time they adopt it, they want it to work properly and to integrate appropriately with their existing technology base. Visionary early adopters and the pragmatist early majority have completely different frame of minds about technology and because of these incompatibilities, early adopter surveys do not help to really understand and to predict accurately how consumer behavior might change as a response to the introduction of the new technology. Findings emerging from the quantitative survey show that the adoption is influenced by perceived interest and it is a function of main value components such as ease of use, perceived usefulness, speed of use, price, and knowledge. These attributes have different importance for each target segment. It is very important that players explore and evaluate the specific needs and requirements of the target segment before launching their services.

The findings have several implications for managers needing to assess the likelihood of success for new service introductions. They also indicate the drivers of acceptance enabling proactive interventions (such as training and marketing activities) to be designed and targeted at populations of users who may be less inclined to adopt and use new systems.

Conjoint analysis revealed "ease of use" as the most significant factor in the adoption process. This illustrates a need for information about digital interactive services usage that marketers can fulfil via alternative channels such as analog television or magazines and the importance of technical and quality usability. Consumers need to be educated about the possibilities of iTV services. An additional factor that will become more prevalent with the introduction of new technologies is price and payment options.

Future research needs to develop a deeper understanding of the dynamic influences studied here, refining measurement of the core constructs used in the model, and understanding the organizational outcomes associated with new technology use.

Investigating Business Models with the Emerging DTV Market

The television landscape is undergoing a period of flux, triggered by the arrival of a number of DTV platforms. On the one hand these include platforms that were also present in the analog market, as terrestrial, satellite, and cable. In addition, there are important new channels including DTV via DSL, WebTV, and mobile TV.

After exploring the value curve, we focus now on emerging business models for digital (interactive) television services.

The critical questions are:

1.   How are the various platform providers using the digital platforms to differentiate themselves from competitors in and across infrastructures?
2.   What differences do we observe in terms of business models used from DTV service providers in some countries in Europe?

Figure 3. Digital iTV services in Europe classified by technology. Base: 1.035 services (Source: New Media&Tv-lab, Bocconi University 2005)

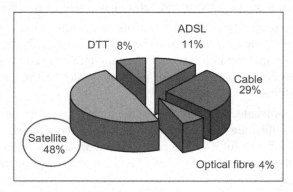

From a methodological point of view, we analyzed 1.035 interactive services provided in Europe (France, Germany, Italy, Spain, UK) by DTV platforms. Each service was analyzed according to three main dimensions: technology, content, and source of revenue.

Forty-eight percent of the services analyzed (mainly enhanced TV and video on demand [VoD]) are broadcast by satellite platforms (Figure 3). Cable and optical fiber allow services with a higher level of interactivity (i.e., interactive advertising, multiplayer games, t-commerce).

Different interactive applications have different profitability levels: VoD, pay per view, home shopping, TV banking, and interactive advertising offer the best potential earnings, while different forms of enhanced TV appear to be less profitable (Table 2).

With reference to the specific content offered Figure 4 shows that iTV services are mainly related to entertainment and information.

The major critical success factor for all the categories of services considered is their content, which makes them attractive face-to-face TV viewers. New ideas and the production of new contents are a must in order to conquer a larger audience, and above all TV viewers must be aware of what is being offered to them.

Interactive services are expected to become an essential feature of DTV services, attract a greater number of subscribers, and create a new source of income and competitive advantage.

The flexibility of digital contents makes it possible not only to offer more and richer traditional services (for example VoD and enhanced TV), but also a wide range of

Table 2. Profitability of different iTV service categories. Source: New Media&TV-lab, Bocconi University 2005

Interactive services	Free P	RTS (Premium Rate Telephony)	Pay-per-service	Transaction
VOD			***	
PPV			***	
Enhanced TV	*	*		
Interactive advertising		*		
Finance		*		
TV shopping		*		***
TV banking		*	*	**
Games, Betting		**	***	**
Communication		*	*	
Web Surfing		*		

new services and applications. These run from enhanced news, VoD, pay per view, to t-commerce, interactive advertising, TV banking, t-medecine, and t-government.

As indicated in Figure 5, different countries in Europe show different strategic orientation concerning the specific interactive applications provided.

In summary, the market development drived by technological innovations, forces broadcasters and digital platforms increasingly to know their positioning and the state of the dynamic competition of the moment.

Figure 4. Digital ITV services in Europe classified by content category. Base 1.035 services. Source: New Media&Tv-lab, Bocconi University 2005

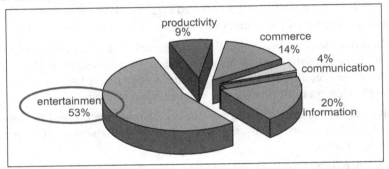

Figure 5. Digital ITV services in Europe classified by content category. Base 1.035 services. Source: New Media&Tv-lab, Bocconi University 2005

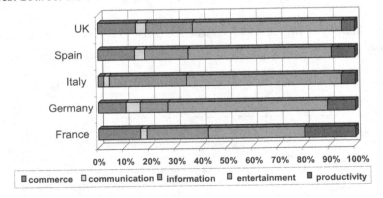

Development Aspects of Business Models for DTV

The technological developments in information and communications technologies have led to a convergence of speech/audio, data, text, graphics, and video and thus to multimedia applications. This field is undergoing rapid development and is set to replace the existing services in the near future (Hulicki, 2005).

Besides, the latest innovations, based on the application of digital techniques, through which the traditional boundaries between media and telecommunications have disappeared, paves also the way for all different kinds of interactive multimedia services.

We can describe the technological and business development in the digital market considering two main dimensions: (1) level of innovation for the broadcaster and (2) level of innovation for the market.

A first development trend is represented by the digitalization of the signal and the launch of new digital TV channels:

- **Digital channels:** Which broadcast without any variation the analog networks' programs.

- **Multiplexing:** Programs are broadcast on different channels in different manners (on different channels the same soccer match is broadcast but with different camera angles).

- **Theme channels:** Specialize on theme sets of programming such as sport, movies, information, or others.

- **Pay per view:** The user pays a fee related to the specific program chosen.

- **Near video on demand (NVoD):** The same broadcast is repeated at very close time intervals (e.g., a movie is broadcast each half hour of the evening on different channels).

Furthermore, a second development driver is offered by digital technology which makes the TV medium more flexible. The user has ample choice of programs, to the point that he/she can create his/her own personalized program set. New iTV services are launched and they include both diffusive numerical services (pay per view, NVoD), and asymmetrical interactive video services (TV banking, TV shopping, interactive games, etc.) (Pagani, 2003).

Also, media integration increases allowing combination of the TV services with Web technologies such as Internet surfing, as well as the carrying out, in the comfort of one's own home, home banking operations, TV shopping, or playing games.

Furthermore, digital technology allows the development of interactive programming and brings existing TV programs to mobile handsets.

Becoming a Customer or Content Gateway in the New Economy

The emerging IDTV marketplace is complex with competing platforms and technologies providing different capabilities and opportunities. iTV has a larger number of key stakeholders and a more complicated set of processes and relationships than traditional TV.

The multi-channel revolution coupled with the developments of interactive technology is truly going to have a profound effect on the supply chain of the TV industry. Interactivity only partially shows the strong innovation in the content industry, in both production and services as well as in operating activities and management styles. Digital technologies have large effects at every stage of the value chain for television broadcasters (from content production to their distribution). These technologies allow for the reduction of production and distribution costs of the television signal, the quality increases but the investment required to purchase content rights are higher. Further, contents will be increasingly important in the emerging market structure.

The competitive development generated by interactivity also creates new business areas, requiring new positioning along the value chain (Porter, 1980) for existing operators.

Figure 6. The network value in Italy

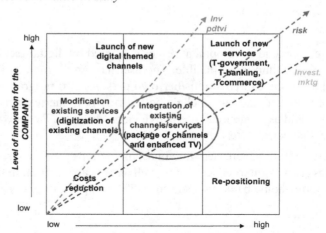

Figure 7.

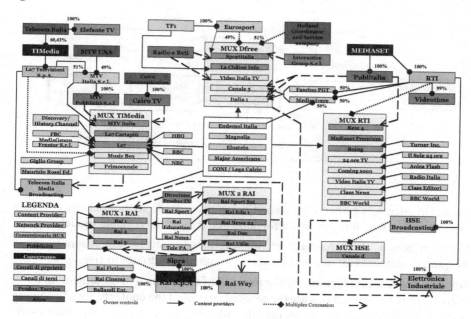

Several types of companies are involved in the IDTV business: content provider; application developer; television producers and broadcasters; network operator; IDTV platform operator; hardware and software developers; Internet developers also interested in developing for television; consultants; research companies; advertising agencies; and so forth.

The *value* chain is underpinned by a particular *value* creating logic and its application results in particular strategic postures. Adopting a *network* perspective provides an alternative perspective that is more suited to new economy organizations, particularly for those where both the product and supply and demand chain are digitized (Peppard & Rylander, 2006).

New business models and revenue opportunities are emerging for operators in the different stages of the value network.

With reference to the flow of revenues among the different operators it emerges that network operators have an advantage in gaining revenue. Their condition of «bottleneck» through which all the content that appears on television must pass, provides a steady revenue stream for pay-TV services while iTV applications are being developed.

Service providers collect their revenues directly from final users through their services. Revenues are then distributed among the other upstream stages of the value

Figure 8. Revenues flow among the value chain elements

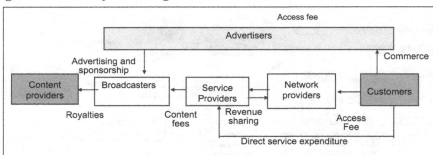

chain. Broadcasters gather relevant additional revenues from indirect sources, such as advertising, which represents a fundamental part of the revenue model.

Within telecoms, the percentage of call revenues funded by advertisers (through freephone, national call rate and local rate numbers) continues to rise.

Some critical success factors, for the purpose of achieving and holding on to long-lasting competitive positions are the gradual control of the contents, thus becoming *content gatekeeper* as well as by the strengthening of the control over final users' access, as *customer gatekeeper*.

Acquiring a position of content and customer gatekeeper gives better chances to determine the ways through which one can develop, assemble, sell, and distribute both the content and services.

Yet, it is worth pointing out that contents do indeed represent the actual killer application in the new DTV market and that hyper competition exists among channels to gain access to contents which most appeal viewers.

In some cases, acquiring control positions for the access of the final users (as a customer gatekeeper) was coupled by a parallel upstream integration of content control leading to the integration of different stages of the value chain of the same entity (e.g., BskyB in UK or RAI in Italy straddle significant portions of the value chain).

The DTV market will be fundamentally more open than its analog forerunner, with intensified competition on the demand side for consumer ownership and on the supply side for content ownership. New business models and forms of commerce will evolve rapidly.

Today's television leaders face a dilemma about the roles they want to play in the DTV landscape of the next 5 to 10 years. Essentially, this boils down to a simple question: do they want to be primarily a world class content owner, or world class consumer owner?

The first option involves developing and exploiting branded content assets through a range of consumer distribution channels and formats.

The second option involves building a business around television-based consumer relationships—so the consumer is king.

Examples of «consumer owners» today include most European cable companies, which look to capitalize on control of the «last mile» and the ultimate direct link with the consumer.

DTV players will often need to decide whether to focus primarily on consumer ownership or content ownership. It will be very difficult to be world class at both.

Vertical integration will be a key supporting strategy for both «content owners» and «consumer owners». Consumer owners can use their own content to build a unique offer, and content owners can use direct consumer relationships to keep in touch with their markets. This is vertical integration that supports a focused consumer or content strategy.

Conclusion: What we can Expect from iDTV

As with all new things, until the consumer has first-hand experience of a medium or product, take-up can only ever be uncertain.

Even with this volume of users, content providers and brand marketers are wondering how they can:

a.   Capitalize on this sizeable audience

b.   Make use of the interactive medium to reach out to and communicate with the audience

c.   Make money online while not cannibalising their core business

A multi-channel environment can further exacerbate increasing audience fragmentation.

DTV is an ongoing reality, and for this reason, building a business strategy which accommodates change is vital.

iTV services are providing new opportunities for brand marketers, keen to pursue closer relationships with a more targeted audience, and with the promise of a new direct sale channel complete with transactional functionality. For broadcasters, garnering marketer support and partners can be a crucial means of reducing costs, providing added bite to marketing DTV to consumers, while establishing new sources of revenue (based on carriage fees from advertisers, revenue shares for transactions coordinated via the DTV platform, and payment for leads generation and data accrued through direct marketing).

At this stage in the iTV market, there are more questions to be answered (Stewart, 1999). The issue at hand are:

- What are the prospects for iTV as a viable medium and (theoretically) in what kind of timescale?
- Can iTV go anyway to alleviating the problem of audience fragmentation? How will the traditional TV advertising model change (will advertisers demand a rate-card based on click through, as with the Web)?
- How can interactivity open up new marketing opportunities (e.g., direct sales channels, direct advertiser-customer marketing relationships, discount-led marketing, leads-generation, data capture)?
- Who will take the lead in developing content for iTV, both in terms of production and creative execution? Will it be broadcasters, who can exact greater control over the environment they actually built? Or will this be agency led?
- Who potentially has the best expertise to develop interactive advertising? Is it the traditional advertising agency, or the interactive specialist agencies, up to now more accustomed to building for the Web? Or will a new form of specialist agency emerge which combines all worldviews? Indeed, might advertisers themselves be able to more directly control these aspects and set up their own production units?

To what extent will the non-interoperability of DTV platforms actually act as a barrier to advertisers embarking on interactive advertising campaigns?

To date, precious few of the aforementioned questions have an inkling of an answer. The number of launch failures for interactive multimedia services in the U.S. and Europe up to now provides an example of the risk rate that is a feature of the new communication markets.

From a strategic point of view, the main concern for broadcasters and advertisers will be how to incorporate the potential for interactivity, maximizing revenue opportunities and avoiding the pitfalls (many unknown) that a brand new medium will afford.

Success will depend upon people's interests for differentiated interactive services. From this chapter three main conclusions can be drawn:

1. Firstly, the development of a clear consumer proposition is crucial in a potential confusing and crowded marketplace.

2.   Secondly, the provision of engaging, or even unique, content will continue to be of prime importance.

3.   Thirdly, the ability to strike the right kind of alliances is a necessity in a climate that is spawning mergers and partnerships.

Those who have developed a coherent strategy for partnering with key companies that can give them distribution and content will naturally be better placed.

Finally, marketing the service and making it attractive to the consumer will require considerable attention, not to mention investment.

References

Afuah, A., & Tucci, C. (2001). *Internet business models and strategies*. Boston, MA: McGraw-Hill.

Amit, R., & Zott, C. (2001). Value creation in e-business. *Strategic Management Journal, 22,* 493-520.

Davis, F. D. (1989). Perceived usefulness, perceived ease of use, and user acceptance of information technology. *MIS Quarterly, 13*(3), 319-340.

Davis, F. D., Bagozzi, R. P., & Warshaw, P. R. (1989). User acceptance of computer technology: A comparison of two theoretical models. *Management Science, 35*(8), 982-1003.

Green, P. E., Helsen, K., & Shandler, B. (1988). Conjoint internal validity under alternative profile presentations. *Journal of Consumer Research, 15,* 392-397.

Hulicki, Z. (2005). Multimedia communication services on digital TV platforms. In M. Pagani (Ed.), *Encyclopedia of multimedia technology and networking* (pp. 678-686). Hershey, PA: Idea Group.

Moore J. F. (1991). *Crossing the chasm.* New York: Harper Business Essentials.

Moore, J. F. (1998). The new corporate form. In D. Tapscott, A. Lowy, & D. Ticoll (Eds), *Blueprint to the digital economy.* New York: McGraw-Hill.

Österle, H. (2000). Enterprise in the information age. In H. Österle, E. Fleisch, & R. Alt (Eds), *Business networking: Shaping enterprise relationships on the Internet.* Berlin, Heidelberg, D: Springer.

Pagani, M. (2000). *Interactive television: A model of analysis of business economic dynamics. JMM Journal of Media Management, 2*(1), 25-37.

Pagani, M. (2003). *Multimedia interactive digital TV: Managing the opportunities created by digital convergence.* Hershey, PA: Idea Group.

Pagani, M. (2004). Determinants of adoption of third generation mobile multimedia services. *Journal of Interactive Marketing, 18*(3).

Peppard, J., & Rylander, A. (2006). From value chain to value network: Insights for mobile operators. *European Management Journal, 24*(2/3), 128-141.

Petrovic, O., Kittl, C., & Teksten, R. D. (2001). Developing business models for e-business. In *Proceedings of International Conference on Electronic Commerce*, Vienna.

Pigneur, Y. (2000). *An ontology for m-business models.* University of Lausanne, Ecole des HEC, CH-1015 Lausanne.

Porter, M. E. (1980). *Competitive strategy.* New York: Free Press.

Rappa, M. (2001). *Managing the digital enterprise—Business models on the Web.* Working paper. North Carolina State University, Raleigh.

Schmid, B. (2000a). Was ist neu an der digitalen Ökonomie? In C. Belz & T. Bieger (Eds), *Dienstleistungskompetenz und innovative Geschäftsmodelle; Forschungsgespräche der Universität St. Gallen.* St. Gallen, Switzerland: Thexis Verlag.

Schmid, B. (2000b). What is new about the digital economy? *EM—Electronic Markets, 11*(1), 44-51.

Stewart, C. (1999). Minefield or goldmine: What can we expect from interactive TV? *New Tv Strategies, 1*(1), 10-14.

Timmers, P. (1998). Business models for electronic commerce. *EM—Electronic Markets*, 8(2), 3-8. Retrieved September 20, 2000, from http://www.electronicmarkets.org/netacademy/publications/all_pk/949 Weill, P., & Vitale, M. R. (2001). *Place to space. Migrating to e-business models.* Boston, MA.

Westland, J. C., & Clark, T. H. K. (1999). *Global electronic commerce: Theory and case studies.* Cambridge, MA: MIT Press.

Endnote

[1] The survey was carried out in 2005 by New Media&Tv-lab (I-LAB Centre for Research on the Digital Economy), Bocconi University.

About the Contributors

George Lekakos is an adjunct lecturer at the Department of Management Science and Technology, Athens University of Economics and Business, Athens, Greece and a visiting lecturer at the Department of Computer Science, University of Cyprus. He holds a BSc in mathematics (Greece), an MSc in software engineering (UK), and a PhD in intelligent media (Greece). His research interests include interactive television applications, digital content management, human-computer interaction, and machine learning with emphasis on the design of personalization algorithms. He has published more than 30 papers in international journals and conferences, and he is the editor of books and conference proceedings. He is a member of the EuroiTV steering committee and heads the Intelligent Media Lab group within the ELTRUN eBusiness Research Center. Since 1999 he has been involved in several European Commission funded research projects and he has been appointed as project reviewer by the European Commission and the General Secretariat of Research and Technology in Greece.

Konstantinos Chorianopoulos is a European Commission Marie Curie fellow at the Department of Architecture at the Bauhaus University of Weimar. He is also an adjunct lecturer at the Department of Product and System Design Engineering at the University of the Aegean. He holds an MEng (electronics and computer engineering, 1999), an MSc (marketing and communication, 2001), and a PhD (human-computer interaction, 2004). During his studies and research, he has been affiliated with engineering, business, and applied arts universities. Since 1997, he has worked in four academic research labs (Greece, UK, Germany), which specialize in the areas of multimedia, e-commerce, intelligent systems and interaction design. He has participated in many EC-funded research projects in the field of human-computer interaction for information, communication and entertainment applications in TV, mobile, and situated computing devices. In 2002, he founded UITV.INFO (http://uitv.info), which is a newsletter and Web portal for interactive television research resources (papers, theses), news and events. He is the main author of more than 10 journal papers and he has lectured internationally (conferences, tutorials, seminars, guest lectures) on several aspects (design, engineering, evaluation, management) of interactive TV. He is serving on the editorial boards of *ACM Computers in Entertainment* and of the *Journal of Virtual Reality and Broadcasting*.

Georgios Doukidis is a professor of information systems in the Department of Management Science and Technology at the Athens University of Economics and Business (AUEB). He holds an MSc and PhD from the London School of Economics, where he taught in the Information Systems Department (1984-90). He is founder and director of the eBusiness Research Center of AUEB (ELTRUN) which is one of the largest in European Business Schools with 30 researchers and specializes on digital TV, supply-chain management, e-business models, digital marketing and IS management. He has published 14 books and more than 200 papers. His two latest books are: *Social and Economic Transformation in the Digital Era* (IGI Global, 2003) and *Consumer Driven Electronic Transformation: Applying New Technologies to Enthuse Consumers* (Springer-Verlag, 2004).

* * *

Ari Ahonen (MSc, Tech; MPsych) has worked 5 years in developing media and mobile services. He has been involved in various projects regarding interactive broadcast services. On the design side, his work ranges from concept design and requirements engineering to interaction design. Another line of work comprises market research and the analysis of novel technology products and services. Currently, Ahonen focuses on applying the human-centered design approach to software development for business-to-business (B2B) markets.

Monica Badella received her degree in computer engineering at Politecnico di Torino in 2001. From March 2001 to June 2005 she worked as an associate researcher inside Applied Pervasive Architecture Lab (ApPeAL) at Politecnico di Torino. She worked on projects related to analysis, evaluation, and design of indoor positioning systems based on Bluetooth technology; service discovery protocols in domestic networks; and architectures and platforms for remote management of terminals. Since June 2005 she has been working as a researcher in the Applications and Services Lab at Istituto Superiore Mario Boella, developing Web applications and services for several projects.

Keith Baker is currently working for Philips Applied Technologies. Royal Philips Electronics is eighth on Fortune's list of global top 30 electronics corporations. Philips is active in about 80 businesses, varying from consumer electronics to medical systems and from security systems to semiconductors. Philips is a world leader in digital technologies for television and display, wireless communications, speech recognition, video compression, storage, and optical products as well as the underlying semiconductor technology that makes these breakthroughs possible.

Andrea Belli was born in Modena, Italy. He graduated in computer engineering from the University of Modena e Reggio Emilia in 2001 with a thesis on content management system based on extensible markup language (XML) technologies. He immediately joined the Telecommunication Research Center for the Telecom Italia Group (formerly CSELT) in Torino, focusing on Internet platforms and cross-media services. Currently, he is involved in a research project which exploits innovative service models about TV consumption over broadcast and broadband connections.

Yolanda Blanco-Fernández was born in Orense, Spain in 1980. She received the Telecommunications Engineering degree from the University of Vigo in 2003. She is currently a PhD student at the Department of Telematics Engineering from the same university, and a student member of the Networking and Software Engineering Research Group. Her major research interests are the design and implementation of services for interactive digital TV (IDTV).

Dick Bulterman is a senior researcher at CWI in Amsterdam, where since 2004 he has led the Distributed Multimedia Languages and Interfaces Theme. From 1988-1994 (and briefly in 2002), he led CWI's Department of Computer Systems and Telematics, and from 1994-1998 he was head of the Multimedia and Human Computer Theme. In 1999, he and two other "brave souls" started Oratrix Development BV, a CWI spin-off company that transferred the group's SMIL-based GRiNS software to many parts of the civilized world. In 2002, after handing the responsibilities of CEO over to Mario Wildvanck, he returned to CWI and started up a new research activity at CWI on distributed multimedia systems. Prior to joining CWI in 1988, he was on the faculty of the Division of Engineering at Brown, where he was part of the Laboratory for Engineering Man/Machine Systems. Other academic appointments include visiting professorships in computer science at Brown (1993-1994), the information theory group at TU Delft (1985), and a part-time appointment in computer science at the University of Utrecht (1989-1991). Bulterman received a PhD in computer science from Brown University (USA) in 1982. He also holds an MSc in computer science from Brown (1977) and a BA in economics from Hope College (1973).

Verolien Cauberghe is a PhD student at the University of Antwerp, Belgium. Her research focuses on IDTV and advertising effectiveness. The impact of interactivity embedded in TV formats on advertising effectiveness is the main angle of her PhD. In this light she set up several experimental studies, whereby she tries to measure advertising effects in an interactive context. Besides this research stream, she looks at the adoption behavior of advertisers on a longitudinal base. Overall, her interest is grasped by new media and new advertising possibilities.

Pablo Cesar is a postdoctoral researcher at the CWI (The National Research Institute for Mathematics and Computer Science in the Netherlands) in Amsterdam. He received a Dr. Tech. degree (December 2005) from Helsinki University of Technology (Finland) and an MSc degree (February 2002) from Universidad Politecnica de Madrid (Spain). The name of his doctoral dissertation is "A Graphics Software Architecture for High-End Interactive TV Terminals." He joined CWI in November 2005 and, currently, he is working on the Passepartout and the BRICKS PD-3 projects, which concentrate on advanced user interaction for digital television. Before joining CWI, he worked on the Future TV (about software architecture for digital television receivers), Otadigi (setting up a digital television broadcast system for the Otaniemi region, Finland), and Brocom (about graphics architecture for multimedia platforms) projects. His research interests include interactive digital media and usability.

Robin Cohen is a professor in the David R. Cheriton School of Computer Science at the University of Waterloo, where she has been a faculty member since 1984. She conducts research in artificial intelligence, in the areas of user modeling, intelligent interaction, and multi-agent systems. She has served on the Advisory Board of User Modeling Inc., has acted as the program chair for UM92 and AI02 (the Canadian AI Conference), and has been on the program committees of numerous conferences and workshops in the artificial intelligence community.

Samuel Cruz-Lara earned a masters degree in computer science in 1984 (University Henri Poincare, Nancy 1) and a PhD in computer science in 1988 (National Polytechnic Institute of Lorraine). The central topic of his PhD thesis was the generation of integrated development environments by using attribute grammars. He is currently associate professor at the University of Nancy 2 (University Institute of Technology, Computer Science Department) and permanent researcher at LORIA/ INRIA Lorraine (Lorraine Laboratory for Research in Computer Science and its Applications—UMR 7503—CNRS—INRIA—Universities of Nancy). He belongs to the Language and Dialogue Team and has conducted several research activities on distributed software architectures, textual linguistic resources management, and multilingual and multimedia resources management. He has participated in several projects, in particular CNRS-SILFIDE, MLIS-ELAN, Digital Museum Project (this project has been sponsored by the National Science Council of Republic of China [Taiwan] numbered NSC-89-2750-P-260-001 and supported by INRIA, France), and ITEA's Jules Verne Project. Currently, he is the project leader at LORIA/INRIA Lorraine of ITEA's Passepartout Project, and he is promoting the definition of multi lingual information framework (MLIF), a high-level, ISO-based abstract model for dealing with multilingual content. In the framework of ITEA's Passepartout project, he is associating MLIF to MPEG and SMIL standards.

Kristof Demeyere was born in Roeselare, Belgium on November 11, 1983. He has a degree in electrical engineering at Ghent Univeristy. In September 2005 he started his final year dissertation on live-event broadcasting and videoconferencing for the multimedia home platform (MHP) at the Department of Information Technology (INTEC), Ghent University (Belgium).

Tom Deryckere was born in Ghent, Belgium on August 25, 1981. He received the MSc degree in electrical engineering from Ghent University (Belgium) in July 2004. In 2004 he joined the Department of Information Technology (INTEC), Ghent University (Belgium) where he is currently working as research assistant in the Wireless and Cable Research Group. He is currently involved in different research projects for the Interdisciplinary Institute for BroadBand Technology (IBBT-Ugent/INTEC) Ghent University. His interests are in the development of interactive television services for fixed and mobile TV, of personalised services and Human Computer Intersection.

Sanaz Fallahkhair is a software engineering graduate of the University of Brighton. She is currently finishing her PhD on language learning via iTV and mobile devices while working as research fellow on the EU project LOGOS, which is developing cross platform learning solutions using archive material.

Marina Geymonat was born in Torino, Italy. She graduated in computer science from the University of Torino in 1993 with a work on network management. In 1994 she joined the Telecommunications Research Center for the Telecom Italia Group. Since then she has been working in different areas, with a strong participation in European projects and standardization activities on asynchronous transfer mode (ATM) technology. She now leads a research project for designing future (iTV), exploiting both the digital terrestrial channel and the broadband connection.

Alberto Gil-Solla was born in Domayo, Spain in 1968. He is an associate professor in the Department of Telematics Engineering at the University of Vigo, and a member of the Networking and Software Engineering Research Group. He received his PhD in computer science from the same university in 2000. Nowadays, he is involved with different aspects of middleware design and interactive multimedia services.

Richard N. Griffiths is a chartered information technology professional and member of the British Computer Society. He has a degree in sociology from Brunel University and a masters in cognitive studies from Sussex University. He first worked for John Hoskins and Company implementing accounting software, then Racal-Hyperon Ltd as a project leader/analyst working on microcomputer applications, systems, and communications software. Subsequently, he taught computing in further education at Brighton College of Technology before joining Brighton Polytechnic, now the University of Brighton as a computing lecturer where he researches and teaches HCI and interactive television topics.

Christoph Haffner has a degree as an archaeologist from the University of Hamburg and worked as an educator in museums. He participated in the development of multimedia productions for archaeological museums and other cultural institutions. He also has a degree in multimedia management. Currently, he is working on IDTV productions and teaches video technologies and iTV as a member of the HCI Research Group at Multimedia Campus Kiel. He is also engaged in multimedia and augmented reality projects for archaeological museums.

Shang Hwa Hsu is a professor in the Department of Industrial Engineering and Management at National Chiao Tung University, Taiwan. He received his PhD degrees in experimental psychology from the University of Georgia in 1980. He has worked as a user interface specialist for Xerox Corporation and Unisys Corporation. His present research interests include HCI, product innovation, and interface design for large-scale complex systems.

Arianna Iatrino received a degree in communication science from the University of Turin in 2005. The title of her thesis was "From Web to Digital Terrestrial Television: The Re-Design of a Tourist Service Called UbiquTO." It concerned the re-design of a Web interface in order to adapt it to the digital terrestrial television context. She was also involved in the UbiquiTO Project in collaboration with the University of Turin. Iatrino is currently working in the CSP Research Centre as expert in design and analysis of MHP applications and interfaces. In particular, she works for the applicative platform area for the development of MHP interfaces for local public administrations.

Michiel Ide was born in Ypres, Belgium, on October 31, 1982. He received the MSc degree in computer science from Ghent University (Belgium) in September 2005. In 2005 he joined the Department of Information Technology (INTEC), Ghent University (Belgium) where he is currently working as a research assistant in the Wireless and Cable Research Group. He is currently involved in different research projects for the Interdisciplinary Institute for BroadBand Technology (IBBT-Ugent/INTEC) Ghent University. His interests are in the development of interactive television services for fixed and mobile TV.

Jens F. Jensen has, since 2001, been professor in interactive multimedia at the Department of Communication, which is a department under the Faculty of Humanities at Aalborg University. Additionally, he is head of the Research Centre for Interactive Digital Media—InDiMedia; head of the Centre for Experience Economy; Creative Industries and Technologies, ExCITe; and head of the Centre for Applied Experience Economy—ApEx. Currently, he is project head for BID-TV (user-centred interactive, digital TV) and Plan B, a project on broadband TV and user-generated content. His primary research interests are digital interactive multimedia, network-based media, media convergence, digital aesthetics, and new media's sociology and user culture.

Jan Kallenbach is currently perusing his PhD at the Helsinki University of Technology (HUT) within the Laboratory of Media Technology. He originated from Ilmenau Technical University in Germany, where he focused on the concepts for the print media-based access to DVB/MHP environment and implementation of a reference application. He is currently part of the Digi-TVtoPrint project at HUT.

Annelies Kaptein works for Stoneroos Interactive Television, a full solution provider for interactive television in the Netherlands. Stoneroos works in close collaboration with programm makers on the one hand and digital TV distribution companies on the other. On a national level Stoneroos, together with her client NCRV, was awarded the Golden SpinAward 2002 for "Best iMedia and iTV concept" for the news show Stand.nl. On an international level Stoneroos is also well-known. In the spring of 2003 Stoneroos won at the MIP/MILIA the Afdesi iTV award for "Best interactive information program," also for Stand.nl. Stoneroos not only won an award in Cannes, but was also nominated for the MoodTV concept in the category "Best un-published iTV concept." Both the award and the nomination emphasize that Stoneroos has struck the right cord in her iTV concepts.

Georgia K. Kastidou is a PhD candidate in the David R. Cheriton School of Computer Science at the University of Waterloo, working under the supervision of Dr. Robin Cohen. She received her MSc degree in 2004 and her BSc degree in 2002 from the University of Ioannina, Greece, both in computer science. Areas of research in which she is interested include multiagent sytems, ubiquitous computing, and mobile computing. The research presented in this chapter was developed as part of the course CS785 Intelligent Computer Interfaces, taught by Professor Robin Cohen in the fall of 2005.

Hendrik Knoche received his diploma in computer science from the University of Hamburg in 1999. He has been working in the field of networked multimedia for 8 years. Currently he is pursuing a PhD in computer science at University College London, researching user experience in mobile multimedia applications with a specific focus on mobile TV and multimedia quality perception.

Heidi Krömker has been a full professor in media production at the Technical University Ilmenau, Germany, since 2002; previously, she was head of the User Interface Design Center of Corporate Technology at Siemens AG. Her team developed innovative user interface designs for industry, energy, communication, information, and medical technology, including research and development projects on augmented reality, intelligent agents, and e-commerce. Working for a global market, the international team runs usability laboratories in China (Beijing), USA (New York/Princeton), and Germany (Munich). After her study of social science she earned her PhD in human factors at the University of Bamberg, Germany.

Tibor Kunert is a research assistant at the Institute of Media Technology of the Technical University of Ilmenau, Germany. He teaches courses on HCI, usability engineering, and media production. His research focus is on user-centered design and development of iTV applications. After his diploma in communication science at the University of Applied Science for Technology and Economics in Berlin, Germany, he worked as concept developer at Pixelpark AG, formerly Germany's largest multimedia agency. There he developed innovative functionality and interaction concepts for several interactive applications, mainly for knowledge management systems, iTV, and electronic banking.

Maria Lahti (MA) works as a usability research scientist at VTT Technical Research Centre of Finland. Her research interests include the usability of digital television and mobile media services as well as the social use of media. She has previously worked as user documentation and usability designer in software development.

Chia-Hoang Lee is a professor in the Department of Computer Science at National Chiao Tung University, Taiwan. Lee received his PhD in computer science from University of Maryland in 1983. He is also the director of MediaTek Research Center in both the Colleges of Electrical Engineering and Computer Science. The center focuses the research in the area of networking, chip design, media, and interface software. He was a faculty member at the University of Maryland and Purdue University from 1984-1992. His current research interests are in the areas of Chinese essay scoring systems, Web services, and man machine interface in mobile devices. Lee served as an associated editor of the *International Journal of Pattern Recognition* in 1998.

Martín López-Nores was born in Pontevedra, Spain in 1980. He received the Telecommunications Engineering Degree from the University of Vigo in 2003. He is currently a PhD student in the Department of Telematics Engineering. His major research interests are the design and development of interactive services for digital TV.

Artur Lugmayr's interest is situated between art and science, where his vision is to create consumer-centered media experiences. He is perusing his second PhD at the School of Motion Picture, TV and Production Design, Helsinki. He is heading the new ambient multimedia (NAMU) research group at the Tampere University of Technology (Finland). He chaired the ISO/IEC ad hoc group "MPEG-21 in Broadcasting"; won the NOKIA Award of 2003; is country representative of the Swan Lake Moving Image & Music Award; and has contributed numerous publications (including a book). He is the inventor of bio-multimedia—integrated human capacity. His passion is filmmaking.

Luc Martens was born in Gent, Belgium on May 14, 1963. He received the MSc degree in electrical engineering and a PhD degree in development of a multi-channel hyperthermia system: electromagnetic modeling of applicators, generator design, and estimation algorithms for thermometry from Ghent University (Belgium), in July 1986 and December 1990, respectively. From September 1986 to December 1990, he was a research assistant at the Department of Information Technology (INTEC), Ghent University. Since January 1991, he has been a member of the permanent staff of the Interuniversity MicroElectronics Center (IMEC), Ghent and is responsible for the research on experimental characterization of the physical layer of telecommunication systems at INTEC. Since April 1993, he has been a professor of electrical applications of electromagnetism at Ghent University.

Sonia Modeo is a PhD student in communication science and design at the University of Turin. She is now working on her PhD thesis concerning the design of an adaptive media center. She is also currently working in the CSP Research Centre as a usability expert. In particular, she is a member of the permanent research group at the W3Lab (http://www.w3lab.csp.it). She received a degree in communication science at the University of Turin in 2003. Her thesis was about the design of an intelligent agent in mobile office application. In 2002 she took a master in Web career at the University of Turin in association with the Getronys society.

Max Mühlhäuser is a full professor of computer science at Darmstadt University of Technology, Germany. He worked as a professor/visiting professor at universities in Germany, Austria, France, Canada, and the U.S. Mühlhäuser published about 200 articles, co-authored and edited books about computer-aided authoring/learning and distributed/multimedia software engineering. His core research interest is development support for next generation Internet applications: ubiquitous, ambient, and mobile computing and commerce; e-learning; multimodal interaction; distributed multimedia and continuous media; hypermedia and Semantic Web; cooperation; and pervasive security. The enabling technologies comprise distributed object-oriented programming, event-based/peer-2-peer infrastructures, hypertext, and audio/video processing (GI, ACM, and IEEE member).

Margherita Pagani is an assistant professor of management at Bocconi University (Milan) and head researcher for New Media&Tv-lab at the I-LAB Centre for Research on the Digital Economy. She is associate editor for the *Journal of Information Science and Technology* (USA). She was visiting scholar at Sloan-MIT (Massachusetts Institute of Technology) and visiting professor at Redlands University (California). She worked with RAI Radiotelevisione Italiana and as a member of the workgroup Digital Terrestrial for the Ministry of Communications in Italy. She is the author of *La Tv nell'era digitale* (EGEA, 2000), *Multimedia and Interactive Digital TV: Managing the Opportunities Created by Digital Convergence* (IRM Press—USA, 2003 and Communication Books—Korea, 2006), *Wireless Technologies in a 3G-4G mobile environment: Exploring new business paradigms*. She has edited the book *Mobile and Wireless Systems Beyond 3G: Managing New Business Opportunities* (IRM Press, 2005) and *Encyclopedia of Multimedia Technology and Networking* (IPG, 2005).

José J. Pazos-Arias was born in Bayona, Spain in 1964. He received his degree in telecommunications engineering from the Polytechnic University of Madrid in 1987 and his PhD in computer science from the Department of Telematics Engineering at the same university in 1995. He is the director of the Networking and Software Engineering Group at the University of Vigo, which is currently involved with projects on middleware and applications for IDTV, receiving funds from both public institutions and industry.

Patrick De Pelsmacker is professor of marketing at the University of Antwerp and the University of Ghent (Belgium). He teaches marketing communications and marketing research. His research focuses on the impact of marketing communications in general and of advertising in particular, and covers issues such as medium context effects; new product strategies and advertising; interactive advertising and nonprofit marketing communications; and consumer behavior. He has co-authored books on marketing communications and marketing research, and published articles in, among others, *Journal of Advertising, International Journal of Research in Marketing, International Marketing Review, Journal of Marketing Communication,* and *International Journal of Advertising.*

Lyn Pemberton is reader in human computer interaction in the School of Computing at the University of Brighton on the English south coast. She has worked for over 15 years in educational software development, particularly on language-related projects. She researches and teaches in HCI, most recently in the use of iTV for learning and teaching. The University of Brighton was responsible for the first EuroITV conference and runs a successful MSc programme in Digital TV Management and Production.

Monica Perrero was born in Chivasso (Torino), Italy. She graduated in communication in ICT Society at University of Torino in 2004 with a thesis on an adaptive location-based service. In 2005 she joined the telecommunication research center for the Telecom Italia Group, where she is working on a research project for designing future iTV, exploiting both the digital terrestrial channel and the broadband connection.

Alexandra Pohl obtained her BA in political science from the Freie Universität Berlin and graduated in 2000 in Audio-Visual Media Studies at the University of Film and Television "Konrad Wolf." Before joining RBB she worked for echtzeit AG and CanalWeb, a French Internet-TV provider where she was responsible for the development and coordination of several interactive live programs. She has been working for RBB's Innovation Projects, contributing to different European funded projects and as RBB's project manager for the SAVANT and the ENTHRONE Projects. In 2004/2005 she coordinated multimedia research and development for RBB's Department of Production and Operation.

Roel Puijk is professor in media studies at Lillehammer University College in Norway. He has mainly written on media events and television production. He was the leader of an international project that studied the mediation of the 1994 Winter Olympics and edited the book *Global Spotlights on Lillehammer* (Luton University Press 1997). Puijk has previously worked at the University of Bergen and the University of Oslo where he conducted an ethnographic study of the Enlightenment Department of the Norwegian Broadcasting Corporation during the 1980s. In 2003 he revisited the corporation to study new production forms. He is currently leader of the research project Television in a Digital Environment (TiDE.hil.no), which is funded by The Research Council of Norway.

Manuel Ramos-Cabrer was born in Lugo, Spain in 1966. He received his degree in telecommunications engineering from Polytechnic University of Madrid (Spain) in 1991, and his PhD degree in Telematics from the University of Vigo (Spain) in 2000. Since 2001, he has been an associate professor in Telematics Engineering at the University of Vigo. His research topics are IDTV concentrating on recommender systems, integration with smart home environments, and interactive applications design and development.

Marjukka Saarijärvi (MSocSc) has worked 10 years in the Research Unit and in the Media Policy Unit of the Ministry of Transport and Communications Finland. Her expertise ranges from information society concepts to service development of public administration. During the years 2004-2005, she was responsible for the Ministry's Digital TV cluster program ArviD. Currently, Saarijärvi works in the local government IT unit of the Ministry of the Interior.

M. Angela Sasse is the professor of human-centered technology in the Department of Computer Science at University College London. Since joining UCL in 1990, she has participated in or led more than a dozen projects on design and usability issues in networked multimedia systems, with a particular focus on assessing media quality and user experience. She has co-authored around 50 peer-reviewed publications in this area. In recent years, her research has focused on quality of experience (QoE) in mobile applications and services. In 2004 and 2005, she chaired the working group on Human Perspective and Service Concepts of the Wireless World Research Forum (http://www.wireless-world-research.org).

Rossana Simeoni was born in Ivrea, Italy. She graduated in computer science at the University of Torino in 1991 with a work on logic programming. In 1992 she joined the Telecommunications Research Center for the Telecom Italia Group. Since then she has been working in different areas, including service management, particularly CRM and service personalization, process engineering, and lately started a research project for designing future iTV, exploiting both the digital terrestrial channel and the broadband connection.

Luiz Fernando Gomes Soares received a DSc in computer science from PUC-Rio (Brazil) in 1983. He also holds an MSc in computer science and an electronic engineering degree from PUC-Rio (1976 and 1979). Soares is a full professor in the Informatics Department in the Catholic University of Rio de Janeiro (PUC-Rio), where since 1990 he heads the TeleMidia Lab. He is a board member of the Brazilian Internet Steering Committee; counselor of the Brazilian Computer Society, and chair of the Media Synchronization Brazilian Consortium for the Brazilian Digital TV System. Prior to joining PUC-Rio he was a researcher at the Brazilian Computer Company. Other academic appointments include visiting professorships in computer science at École Nationale Superieure de Télécomunications (France), Université Blaise Pascal (France), and Universidad Federico Santa Maria (Chile). He also spent 2 years at the IBM Scientific Center in Rio.

Mark V. Springett is a senior lecturer at Middlesex University, UK. He teaches HCI design at postgraduate level and is responsible for the development of distance-learning provision for postgraduate courses. He is a member of the Interaction Design Centre and the Design-for-All research group at Middlesex. He is a member of the BCS/HCI. He is on the management committee of MAUSE COST action 194, 'Towards the Maturation of Information Technology Usability Evaluation'. Springett gained his PhD in HCI design at City University, then he took a lectureship at the University of North London (now London Metropolitan University). He worked as a Usability Consultant with Openwave Systems before taking his current post.

Raivo Suni holds an MA (1976) in media studies from the University of Oslo (Norway). Currently he is a PhD student of journalism and communication at the University of Tartu (Estonia). In his recent research, Suni has focused on audience studies of Internet protocol television (IPTV). In addition to his studies, Suni works as an audience analyst and program consultant for the Estonian Public Broadcasting Television Channel (ETV). Suni is also a member of the Baltic Association for Media Research (BAMR) and Estonian Press Council.

Laura Turkki (MSc, Tech) has worked in all phases of user-centered product development with a wide variety of products, ranging from medical equipment to online games and digital television services. She has previously worked for Helsinki University of Technology; Adage, a usability consulting company; and she currently works at Nordea Netbanks.

Tytti Virtanen, an MSocSc student, works as a trainee research scientist at VTT Technical Research Centre of Finland. Virtanen's area of interest is human-technology interaction focusing on digital television and mobile usability, and she is currently doing research on user experience measurement.

Thorsten Völkel received a diploma degree in computer science from the Christian-Albrechts University of Kiel. After his study he joined the Human-Centered Interfaces Research Group of Professor Gerhard Weber at the Multimedia Campus Kiel, where he currently works as a research and teaching assistant. His main research interests include the design process of multimodal fusion in conjunction with aspects of accessibility. Völkel is author and co-author of several conference papers in the areas of formal modeling of multimodal fusion as well as accessibility of mobile devices.

Ming-Hui Wen is currently a PhD student in the Department of Industrial Engineering and Management at National Chiao Tung University, Taiwan. His present research interests include user interface design and product innovation.

Index